instant evolution

instant evolution

we'd better get good at it

Thomas P. Carney

UNIVERSITY OF NOTRE DAME PRESS
NOTRE DAME LONDON

Notre Dame, Indiana 46556

Library of Congress Cataloging in Publication Data

Carney, Thomas P
Instant evolution.

Includes index.
1. Bioethics. 2. Medical ethics. 3. Biology—
Social aspects. 4. Social medicine. I. Title.
QH332.C37 174′.2 79-17835
ISBN 0-268-01145-1

Manufactured in the United States of America

To
Dr. George N. Shuster
who made this book probable

contents

Introduction ix

Prologue 1

1 . . . wherein the scientist goes from self-satisfaction and isolation to public awareness 3

2 . . . wherein life begins and life ends and it is not known when, and life can be continued and life can be eliminated and it is not known that it should be 10

3 . . . wherein many and diverse genetic defects are explained, and they exist in good and bad and obvious and hidden forms 39

4 . . . wherein it becomes possible to change the blueprint of life 53

5 . . . wherein a conflict develops between the ability to detect genetic defects and the desirability of such detection 74

6 . . . wherein genetic variations occur in every human being, and some genes or individuals should be eliminated, changed, or tolerated 87

7 . . . wherein control of reproduction and the development of new reproductive methods become realities 98

8 . . . wherein composite human beings are made and their minds controlled 115

9 . . . wherein human beings are subjects for experimentation, and the rights of individuals are contrasted to the rights of society 125

10 . . . wherein it is shown that we are as gods, and now need to learn to act as gods 149

Epilogue 169

Notes 171

Index 175

introduction

This is a book about science and medicine, scientists and physicians, ethics and morals, and the problems caused for some or all of us by some or all of them.

It is not written for a specialized audience. I hope it will be read by those who think that science, in the form of arrogant scientists, has no compassion for the eventual legatees of laboratory results. I hope it will be read by those who believe that science is, of itself, an end, and has no limits on what can be done in its name in the search for truth. I hope it will be read by those interested in learning of the ethical and moral problems that have been created by the almost unbelievable increase in basic and applied knowledge resulting from research. I hope it will be read by those who think that science is responsible for our problems and must, therefore, be prevented from further research, as well as by those who think that all we need to solve the problems is more science.

Certainly we are faced with ethical dilemmas not anticipated even a decade ago. Many of our problems became apparent as the result of a new consciousness of the right of the individual. No longer is the dominant consideration the common good. In human experimentation, for example, each individual taking part in an experiment must give truly informed consent. He or she must be aware of the risks involved. An experiment cannot be justified on the basis that "A few will be hurt, but the results will benefit many." If that is to be the ultimate result, it must be accomplished with the approval of each individual involved.

Ethical considerations have always been a part of research. When research can affect little, the problems are minor. When research results in drastic change, the problems are great.

I once believed that science of itself was sufficient. I believed that the scientist should obtain results, and that someone else should be responsible for how the results were used. I believed that scientists had no responsibility for the uses that were made of their work, and I believed that scientists should be able to do any research that they were free to do, without interference from non-scientists.

The subtitle of this book was inspired by a paragraph on the first page of *The Whole Earth Catalog;* it says: "We are as gods, and might as well get good at it." I feel more strongly than that. We'd *better* get good at it if we want to survive.

"We are as gods." Gods decide who will work and who will be idle, who will be fed and who will go hungry. They decide who will stay sick, and who will be cured, who will live and who will die. They even decide who will be born and who will not be born.

In the following pages I discuss some of the problems facing us today. I have felt it necessary to include a more technical description of some things than might be desirable in a book of this kind. However, I believe the basic understanding of the background of, for example, the propagation of genetic traits is essential if we are to expect to evaluate alternatives in these areas.

The responsibility for decisions no longer rests with only one group, whether it be scientists, ethicists, theologians, or various levels of government. These problems are the concerns of everyone. It is the responsibility of the scientist not just to inform but to educate those not in science as to the possible consequences of the results of research. In turn, it is the responsibility of the public to be informed and to learn, so that in cooperation with the scientists, rational, rather than emotional, considerations can be used to arrive at decisions.

The difficulty we now face in rationalizing the differences between science and ethics arises from the different concerns of the two. Science is concerned with what *is,* ethics with what *should be*. If we want to rationalize the two by starting with the idea that science is predominant, and anything it does is good, then we can only say that "What is should be," and the ethical considerations must be changed as new scientific results are obtained. However, ethics is also concerned with good and bad, right and wrong, and cannot say that what is good now is bad tomorrow because of a scientific result. Therefore, the ideal rationalization brings us to the conclusion that science acts in such a way that "What should be is."

How we solve our ethical, moral, and scientific problems will determine what humanity will be in the future.

We'd better get good at it.

Thomas P. Carney

prologue

To the casual observer everything seemed to be going along smoothly at the plant. The ova collectors had done a good job of building up a group of women with the proper genetic composition to act as ova donors. Some donors had been given chemicals called superovulators, resulting in the production of numerous ova every month. Efficiency over the last quarter had increased significantly.

With the new method developed in research the Quality Control Department was now able to detect chromosomal abnormalities immediately after fertilization in the test tube, so that it was no longer necessary to waste synthetic uterus space on embryos that would eventually be rejected. Early elimination of rejects also made it easier for the Distribution Department to plan for the priority delivery of fertilized eggs to the surrogate farm for implantation into women who would carry them to term. Those fertilized eggs to be transferred from the test tube to the synthetic uterus did not present the same urgent problem, since it did not make too much difference whether or not their degree of development was carefully controlled.

Of course the Special Services Department was a source of irritation. The problem of fertilizing a wife's ovum with her own husband's sperm caused no end of paper work and special control. There still seemed to exist a small number of fundamentalists who believed they should have some control over the offspring they produced. The time did not yet appear appropriate to eliminate such independent and inefficient operation.

But things were not all as smooth as they seemed. A recall had been ordered on a full month's baby production. A check of the records showed that an operator, in an outdated expression of compassion for some of the embryos that would have been discarded under present policy, had changed the conditions in the synthetic uterus. As a result, all the babies in that group developed a kidney defect that did not become apparent until after a year of growth. Such recalls caused great agitation, since it was no easy matter to have a large volume of kidney transplants performed in a short period of time. In addition, it was a considerable strain on the body banks, since the resuscitators would now be working on kidneyless bodies until a new crop of embryos could be developed. It was incidents such as this that gave ammunition to those of the Button Pushers who had been agitating for some time to have kidneys replaced with the newly developed plastic organ as soon as the fetus came out of the incubating uterus.

The Euthanasia Section had not escaped its share of problems. The demand for its products—the organs of those subjected to the "accelerated transfer from uncomfortable living to a peaceful rest" procedure—had been decreasing rapidly, since, with the design of artificial organs and the success of the embryo banks it was no longer necessary to use them as a source of transplants. Some of the new employees were even beginning to question whether or not the elderly and the unfit should be eliminated. Their lack of understanding of the problem was reflected in the more frequent use of the phrase "put to death" instead of the more euphemistic and customary expression "relieve them of their troubles."

Much more serious was the report that the first large-scale attempt to change the potential of individuals by modifying their DNA patterns had not been a complete success. Genetic patterns had been changed as predicted, but the resulting individuals did not respond in the uniform way that had been desired. The problem now being faced was what to do with these people who did not fit into any of the established behavioral categories.

All in all it had not been a good year for the Humanity Modifiers.

1 . . . wherein the scientist goes from self-satisfaction and isolation to public awareness

It was a good time to be involved in medical research, the 1930s, '40s, and '50s. It was a time when miracles became routine. The sulpha drugs, the antibiotics, the antihistamines, the tranquilizers, tissue culture, polio vaccine, analgesics, oral contraceptives—no other age had ever produced, nor will a future age ever again produce, such a litany of benefits for human society. It was an exciting time, a stimulating time.

I was a part of that time. I never had any desire to be a physician, to diagnose diseases, to administer drugs to patients, to perform operations, but I have been fiercely proud to be a part of a profession that could conceive and develop and make available drugs to be used by others for the relief of human suffering.

Back in the middle of the seventeenth century a chemist named Becher wrote what has been called the Chemists' Creed:

> The chymists are a strange class of mortals, impelled by almost insane impulse to seek their pleasure among smoke and vapors, soot and flame, poisons and poverty. Yet, among these evils I seem to live so sweetly that may I die if I would change places with the Persian king.

When I see what has resulted from my laboratories I, too, would not change places with the Persian king.

It was a comfortable time, that time of scientific miracles. The scientist had only one objective then: to do research, to get results. If the work was in basic research, satisfaction came from the doing of the work. If the scientists were in applied research, satisfaction came both from the work and from seeing someone else use the results to produce a useful drug.

My own first experience with this kind of satisfaction came very early in my career, while I was still doing postdoctoral work in a university. I had been studying topical anesthetics, drugs that, when applied to a surface wound such as a burn, would relieve pain. I was successful in this research, and I shall never forget how I felt when I saw this drug used for the first time on a three-year-old girl who had been scalded with boiling water. Within minutes her pain was relieved. It is difficult to describe the feeling of knowing that this was the result of something I had conceived with my own mind and made with my own hands.

My first job in an industrial research organization after completing my doctoral studies was with a company already known for its research abilities. It was a group that, in the early 1920s, recognized the significance of a basic research result on the use of insulin, and had used this information as a justification for producing insulin in commercial quantities and making it available to the world's diabetics. The same foresight was used in making liver extract available for the treatment of anemia. These two advances alone have been responsible for saving millions of lives, and bringing to millions of others relief from symptoms that would have made living intolerable. Both the basic work on insulin and the basic work on liver extract resulted in Nobel prizes for the original researchers.

The same research group had pioneered the work on the barbiturates, and these were already commercially available products. So it was into an atmosphere of scientific triumph that I was fortunate to be thrown.

This was just at the beginning of the antibiotic era. I can well remember the buildings filled with tens of thousands of individual one-litre bottles, each growing a tiny amount of penicillin that would eventually be harvested for use. I can even remember the first patient exposed to our penicillin. The product itself contained no more than four or five percent of the active drug (the rest being impurities), but it *was* active, and the situation involved a lifesaving crisis. The product worked, but the patient expressed the opinion that he would prefer to die rather than repeat suffering the irritation caused by the impurities. Now, with the drug regulations today, it would not have been possible, for legal reasons, to have saved that life. Not long after, as a result of a quantum advance in knowledge of penicillin growth coming out of our laboratories, the one-litre flask gave way to

400,000-litre tanks, and, again as the result of work from our laboratories, entirely new penicillins were developed and produced.

In subsequent years our laboratories conceived and developed the first non-addicting analgesic, antihistiminic agents, tranquilizers, and the first product that was a cure for a specific type of cancer, choriocarcinoma.

The field of basic research was not neglected. Work was just being initiated in academic and government laboratories, specifically the National Institutes of Health (NIH), on the technique of tissue culture. By means of this process, various animal cells, including human cells, could be grown in test tubes using an artificial medium. We had no known use for this technique, but the mere concept of being able to grow human cells outside of the body was exciting enough to justify establishing the first tissue culture laboratory in an industrial research organization.

A couple of years later Salk polio vaccine came along. The basis for the process by which the vaccine is made was the growing of polio virus on animal cells in tissue culture. Thus, because we had established our laboratory as a basic research tool, we were in an ideal position to take Dr. Salk's product and make it practical.

It turned out to be one of the most dramatic programs in the whole history of medicine. It also turned out to be one of the most traumatic. Several laboratories attempted to produce the product. Some of the material had been made and tested in children. Results were positive, and there was general euphoria as to the ultimate result.

Production of the vaccine was proceeding—then tragedy struck. I was driving my daughter to a music lesson and listening to the evening news on my car radio. The lead story reported that a number of cases of polio had been discovered in children who had taken the vaccine produced by another company for trial. My reaction was immediate and instinctive. I made a U-turn on a three-lane one-way street, and headed for home where I could get more details. An understanding police officer allowed me to proceed when I explained the situation.

It was a critical situation. Polio vaccine is made by growing polio virus on tissue culture, and then killing the virus. Obviously, if the virus is not killed, the recipient of the vaccine will be infected by the disease. This is what had happened. Inadvertently, some live virus remained undetected in the vaccine, and some of the children came down with the disease. Because of all the publicity the polio vaccine program had received, this incident could have killed the entire program. However, since our laboratory had so much experience with tissue culture, and because we had the best people in the world working on the problem, we were able to show that there was no danger from *our* vaccine. The program was saved, and for about two years we were the source of the product that eliminated a dread scourge from the country. The fact that attenuated live vaccine, because of its convenience, has replaced the Salk vaccine is almost incidental. With or without live vaccine, polio would have been eliminated.

While working in a different laboratory, I had the opportunity to be associated with another problem that has had a revolutionary effect—oral contraceptives. The antiovulatory, or contraceptive, properties of these products ironically were discovered during a study on the effects of these materials on *increasing* fertility. Laboratories all over the world had been working for years on what are now the constituents of the oral contraceptives—estrogens and progestins. These materials could have been introduced into general practice several years before they finally became available if pharmaceutical companies had been willing to accept the consequences of what they predicted would be adverse public relations reactions. The reactions feared were not those that we would anticipate today, questions about the effect of contraceptives on the whole social system. Rather, the reluctance to enter this field first stemmed from a belief that there would be a vigorous reaction from the Catholic church, not the least result of which would have been a great loss of business to the pharmaceutical company from the hospitals controlled by religious groups.

In 1965, many years before the public apprehension about recombinant DNA, our laboratory abroad was stripping the coats from virus and discussing the possibility of using modified virus to cure disease.

Those were comfortable times. But they are gone forever.

In those days, when we talked about risk/benefit ratios—and we did talk about them a lot—our discussions were limited to what toxic effects might appear in a patient compared to the beneficial effect of the drug. There was not much talk about the social consequences of new discoveries from research, because there was no perspective from which such discussions could take off. The barbiturates were available, but, aside from the possible overuse to aid in sleep, they were not the problem they are today. Some vaccines were available, insulin had been in use for some years, but the implications of the use of these products were considered rather limited.

From a medical viewpoint, the value of the sulfa drugs and penicillin was recognized very early. It was apparent that many lives could be saved that formerly would have been lost. However, the social implication of dramatically influencing the death rate of infants and children was not recognized. Where formerly death controlled the size of families by eliminating the weak, family size was now increasing without an increase in the birth rate.

The tranquilizers came along early in this period. They were truly a giant step forward in medicine. But they soon came to be used indiscriminately by those seeking not relief from an affliction, but relief from normal daily stress and responsibility.

Lysergic acid diethylamide (LSD) was investigated thoroughly as a possible prototype for one mental disease, and as a cure or treatment for others. Its abuse is too well-known to need comment. Then came the oral contraceptives with the real and obvious potential for affecting both social attitudes and social structures. And now with the era of genetic knowledge and genetic manipulation we have come to the point where the social effects can be considered more significant than the medical effects.

Certainly the scientists were not raising questions. If any small doubts ever did surface they were avoided by falling back on the attitude that scientists were not responsible for the way in which the results of their work would be used. Why should they raise questions? What harm could come from the miracles of barbiturates and tranquilizers and antibiotics and contraceptives?

Now we know. But should the scientist and the public have known in the 1920s that the barbiturates would someday become a social problem? I think not. In the early days there was no historical perspective to warn that unpredictable results might arise from the application of new science or technology. In the past, too, even if risk had been predictable, it might have been easy to justify taking since we were usually starting from a zero base. If there is no solution to a critical problem, such as communicable diseases, a solution such as penicillin would be worth trying, almost regardless of the risk involved. However, once the first miracle had occurred, that then became the base. To go from zero to a miracle obviously justifies more risk than going from a miracle to even another miracle.

Today we do have a historical perspective that tells us that practically every result of research can be used in a technology that can be either helpful or harmful, that can be either good or evil. With that recognition comes a whole new level of responsibility both for the scientist and society.

I was a part of that medical revolution when the time was exciting and stimulating and comfortable. I saw the products coming out of our laboratories, and I thought the world was wonderful, and everything in it. Today the times are still stimulating and exciting, but they are no longer as comfortable as they once were.

The last ten years have not produced the products of the 1930s, '40s and '50s, but they have produced techniques whose uses could have an even more revolutionary effect on the human race than have the medical products already available. The last ten years have also seen the development in society of an awareness of the consequences of science and technology. And along with that awareness has come a questioning of the responsibilities of one individual to another, of the rights of society and the rights of the individual, of the protection of the rights of society while preserving the dignity of the individual. And doubts have been raised, not only about how the research information should be used, but even about the advisability of obtaining certain kinds of information.

I too once believed that it would be sufficient to pursue science for its own sake. I too found it inconceivable that anyone could believe that the result of scientific research would be anything but beneficial. Science must progress. If the results of his or her research were misused, it was no concern of the scientist. I too once believed that society should have no influence on how technology developed or how scientific results were obtained.

I no longer believe this. Technology can no longer divorce itself from value judgments. Science of itself may be considered amoral, but the application of its results through technology cannot be. Possibly it is good that as science and technology are recognized as an indispensable part of our present and our future, it should also be recognized that the age of technological innocence has passed. The fact that we can interfere in nature's processes in ways only dreamed of even a generation ago has created moral and ethical problems of the first magnitude. The problems are no longer those of only the scientist. It is the right and the obligation of every individual to be informed of, even to be a part of giving direction to, the sometimes irreversible effects resulting from our actions. In some instances, what we do will affect not only those living today but our children and our children's children far into the future.

2 . . . *wherein life begins and life ends and it is not known when, and life can be continued and life can be eliminated and it is not known that it should be*

In the light of the developments in medicine over the last forty years, it is difficult to imagine even in retrospect the simplistic attitude that prevailed in the laboratories of those early days. We thought that the drugs and techniques we were developing would be used to save lives or relieve suffering. When they did our objective had been reached, and we were happy.

We did not have to decide whether a life would be taken or saved. Lives were for saving. Abortion, euthanasia—these were the realm of the theologian, the philosopher. It did not occur to the scientist that the result of his or her work might make possible a choice between living and dying, even between being born and not being born. Nor did we spend much time thinking about who was alive and who was dead, nor who was human and what was not human. We already knew.

Now science and technology have made it not only possible but necessary to ask these questions.

All over the world human beings are deciding what other human beings will live or die; who will be born and who will not be born. Decisions are made as to who will be allowed to be ill and who will have their illnesses eliminated, whose standards of life will be improved by the provision of food and other necessities, and who will be allowed to exist in misery. When food that might be made available is withheld, a decision in favor of misery, even death, has been made. When there are two candidates for the same kidney machine, a decision has to be made between them, allowing one to live and one to die. When a person unable to maintain his or her normal physiological function is placed on a resuscitator, a decision in favor of life has been made. When abortion is elected, it has been decided the fetus has no right to a life of its own.

There is probably no more important decision that can be made than one involving the life or death of an individual. How these decisions are made constitutes one of the gravest ethical problems of our time. It is grave because the decisions involve human beings. If a living person is judged not to be human, the decision is simplified.

It can be asked why it is important to decide who is human when also deciding who may be killed and who must be protected. The reason must be based in the Western religious and social belief that all human beings are superior to all other living things, and that human beings may use any non-human thing in any way they see fit, even to killing it, as long as the use itself does not dehumanize the human being.

St. Thomas expresses this opinion in his discussion of the treatment of animals. St. Augustine, too, says that the killing of animals and plants is justified since they have no faculty of reason and that "both their life and their death are a matter subordinate to our needs." Slavery around the world was justified on the basis that the victims were less than human.

One of the most famous cases involving a choice between life and death is recorded in legal records as *U.S.* vs. *Holmes*. In 1841 an American ship, the *William Brown,* hit an iceberg near Newfoundland on a trip from Liverpool to Philadelphia. The crew and some of the passengers managed to launch two lifeboats. However, one of these began to sink because it was leaking and overcrowded. In an effort to save some of the people in the boat, the crew threw 14 men overboard, using as a method of selection only the criteria that they would "not part husband and wife and not to throw over any women." Those remaining in the boat were rescued in a matter of hours. Only one of the crew, Holmes, who was acting on orders from the mate during the episode, was found later in Philadelphia. He was convicted of unlawful homicide.

Reactions to this case span the entire range of possibilities for solution. The judge based his decision on his belief that lots should have been drawn to see who should be thrown overboard, since no other action is "so consonant both to humanity and to justice." Others suggested that the selection be based on utility. Those should survive who would be most "useful."

At the other extreme, Edmond Cahn in *The Moral Decision* believed, "If none sacrifice themselves of free will to spare the others—they must all wait and die together."[1]

Nor are life and death decisions restricted to dramatic episodes like a disaster at sea. During World War II penicillin was introduced to battlefield medicine for the first time. It was in very scarce supply, so considerable judgment was used in deciding how best to obtain its maximum benefit. In a choice between giving the drug to soldiers wounded in battle or those suffering from venereal disease, the officers chose to treat the venereal disease first. After all, these soldiers would be returned to the fighting front much faster than those who were wounded.

Let us take another example. One of the most important problems of our age is concerned with what assistance the United States should give to developing nations. It has been suggested that we cannot support, that we do not have the resources to support, all of these countries and, therefore, some process must be developed to aid in deciding where our help should go. The military-developed triage would be one method—some who would benefit from help would be aided in surviving, some would survive without our help, and the others would be allowed to die.

In justifying this type of decision Garrett Hardin proposed what has come to be known as the "Lifeboat Ethic." Hardin rejects Kenneth Boulding's metaphor of "spaceship earth." The metaphor breaks down, he says, because earth has no single captain with decision making authority. Instead, the nations of the earth are more like lifeboats floating on the sea: "Each rich nation amounts to a lifeboat full of comparatively rich people. The poor of the world are in other, more crowded lifeboats. Continuously, so to speak, the poor fall out of their lifeboats and swim for a while in the water outside, hoping to be admitted to a rich lifeboat or in some other way to benefit from the single 'goodies' on board."[2] Hardin argues that if the United States tried to take everyone aboard our lifeboat it would eventually sink and everyone would be lost. Some should be allowed to die now, he concludes, since they will eventually die in any case; by elimination now they will not jeopardize the future of others.

Paul Ehrlich, in *The Population Bomb,*[3] followed this same argument as did Dr. Phillip Handler, president of the National Academy of Sciences.

Amnon Goldworth, of the University of California at Berkeley, has modified both the "spaceship earth" and the "lifeboat" metaphors by suggesting that the world is a sea on which a single, huge ship floats. On the top deck of the ship are the rich, on the lower deck are the poor. The rich are free to engage in sports, in cultural activities, in all those things that improve their physical and mental well-being. But, as Goldworth says, this is due in part to the contribution of those below deck. They work incessantly, with little opportunity to enjoy the privileges of those on the upper deck. They feed the boilers that provide power, and, in general, do all the menial things necessary to keep the boat afloat. Their reward is existence. Their one pleasure is in sex, and this, because of ignorance, tradition, and superstition increases the population of the lower deck to bursting.

Goldworth has used the metaphor to emphasize the responsibility the rich have to the poor. He includes the question of who owns the earth's resources, and implies the interdependence of all the planet's inhabitants.[4]

Daniel Callahan of the Society for Ethics and Life Sciences also seeks alternatives to triage and the lifeboat ethic. He says that while we do have a responsibility to future generations, the more immediate obligation is to the living. He points out that one of the divisions of triage is that of the candidates who cannot live, even with help, and says that there is no firm evidence to sustain a thesis that any of the poor countries is in so hopeless a condition that it must be written off. He therefore suggests that we act as if each can be saved.[5]

What does foreign aid, or military triage, or a shipwreck 150 years ago have to do with biomedical ethics? Probably nothing directly. But our reaction to the situations outlined above is a reflection of the human sensibilities we can bring to bear on the exposure of individuals or even populations to new biomedical techniques. There should be no difference between our reaction to an Indian who is starving and to our neighbor who has been denied the use of a kidney machine. The starving populations, dying soldiers, and shipwrecked voyagers are unknown and distant. It is sometimes difficult to realize the importance of issues when the people affected are far away. Considering these dilemmas as affecting an individual person, especially one we know and love, sharpens our sensibilities and makes us more critical of procedures.

Life

Who cares about the exact time of the beginning of life? Who cares about the exact moment of death? Until relatively recently these questions remained the domain of the theologian and the philosopher, a few lawyers interested in who survived whom, and occasional hospital public relations persons who wanted the first birth of the new year to take place in their institution. Human beings were born and died, and knowing the exact moment this happened had about the same relevance to daily living as did the knowledge of how many angels could dance on the head of a pin. But all that has changed.

The most explosive issue in bio-manipulations, and the one that holds the most implications for a definition of life and death, is abortion, and peripherally, fetal experimentation. And all of that hangs very heavily on the exact moment a life becomes human.

California, North Carolina, and Colorado were the first states to make abortion easier to obtain. By 1970, three years later, fifteen states had adopted similar laws. In 1970 about 200,000 women in the United States obtained legal abortions. In 1971 the number had risen to about 500,000. In each year about 60 percent of the patients were under the age of twenty-five, and 67 percent were unmarried. In 1973 the Supreme Court decided that abortion on demand was legal during the first six months of pregnancy in all the states. In the following three years the number of abortions rose to well over one million, with 300,000 of these being paid for by Medicaid. In Washington, D.C. in 1973 there were 2.35 abortions for every live birth. During the same year in New York there was almost 1 (actually 0.9) abortion for every live birth. In 1974 the figures for the State of New York were 0.698 abortions per live birth, for New York City, 1.14 and for Washington, 1.6. In its journal, *People,* the International Planned Parenthood Federation estimates that between 30 and 50 million abortions are performed around the world each year. This means that there is one abortion for every three live births.

The battle is joined between those who favor abortion as the right of a woman to control her own body, and those who oppose abortion because they regard the destruction of a fetus as murder. The word "battle" is used advisedly. The anti-abortionists, under the "Right to Life" banner, have mounted a vigorous campaign aimed at legislative groups and the general public. Efforts are being made to prevent the use of federal or state funds for purposes of carrying out abortion. Candidates for public office are supported or opposed on the basis of their views on abortion. The topic became an issue in the presidential election of 1976. Following the Supreme Court decision approving abortion on demand, efforts have been made to adopt a constitutional amendment overturning the Court's decision. Public demonstrations, reminiscent of the student activism in the 1960s, are carried out by both pro- and anti-abortion forces. Confrontations between the two groups have resulted in bitter disagreements.

The public atmosphere has been such that one physician, Dr. Kenneth Edelin, the Chief Resident in Obstetrics and Gynecology at Boston City Hospital, was indicted for manslaughter in connection with the death of a fetus during an abortion. The language of the Grand Jury that brought in the indictment was unusual. It said that the physician "did assault and beat a certain person, to wit: a male child described to said jurors as Baby Boy and by such assault and beating did kill the said person." In 1974 Edelin was convicted of manslaughter. Three years later the Massachusetts Supreme Court reversed the conviction in a unanimous decision.

The national embarrassment over situations made possible by the legalization of abortion continues as more moral and legal inconsistencies become known.

In February of 1978 a physician went on trial for the strangling of a baby girl born after an unsuccessful abortion attempt at Westminster Community Hospital in Orange County, California. The abortion, performed on March 2, 1977, involved an eighteen-year-old woman who was seven-and-a-half months pregnant. Twelve hours after saline injection, a baby was born alive. Resuscitative care was started by the medical personnel, but when the attending physician heard of the survival he ordered the procedure stopped. Another physician testified that he had seen the physician in charge apply pressure to the baby's neck on four occasions until no respiration or heartbeat could be detected.

The cause of death was listed officially as asphyxia due to manual strangulation. The defense rested on the fact that the 1973 ruling of the Supreme Court legalized the killing of unborn children, and, therefore, the physician could not be charged with murder since he was only completing what was legal.

The trial lasted fourteen weeks. After eleven days of delibera-

tion the jury reported that it was deadlocked at 7–5, with the majority favoring acquittal, and a mistrial was declared. Comments by individual jurors, even those who had voted for conviction, indicated that they did not believe the physician strangled the baby. However, they doubted that he had tried hard enough to save a life. They were judging him guilty of "omission."

In a related case, the Supreme Court has agreed to rule on the constitutionality of a Pennsylvania law that requires that during the second trimester of pregnancy, when an infant is born who may be medically "viable," i.e., able to survive, abortion must be performed in a way that would "provide the best opportunity for the fetus to be aborted alive."

It sounds like madness—to be required to use a procedure for killing a fetus that would give the best chance for the fetus to be born alive!

What is the purpose and result of such thinking? At best, it would relieve the mother of several months of pregnancy. It would probably require the expenditure of thousands of dollars of premature care in the hospital for the infant. But it would also result in the probability that such infants would all be labelled "high risk," susceptible to brain damage and other handicaps associated with prematurity. It is hard to imagine any great number of infants surviving a saline abortion without damage. The next step in the whole illogical procedure would be to suggest that, even though viable, the life should be destroyed because the procedure used for abortion had caused irreparable damage. It would correspond to the situation of a man who had murdered his parents throwing himself on the mercy of the courts because he has become an orphan.

The Pennsylvania law thus makes it perfectly legal for a physician to try to kill a fetus one minute, but in the next minute, after the fetus has been born alive, the law also requires the doctor to make every effort to save the life he or she had just been trying to eliminate.

I can only ascribe reactions that result in such laws to embarrassment, discomfort, even guilt. The abortion of millions of fetuses has not fulfilled its anticipated purpose. A price is being paid for the taking of all those lives. The cost will not be eliminated by the conscience-hiding device of saying "abort but save the lives." It won't work, and sooner or later the real question will have to be faced—is a fetus a person or not?

The abortion rhetoric on both sides of the issue leaves much to be desired. All those in favor of abortion certainly are not murderers, as the anti-abortionists shrilly insist. On the other hand the claimed benefits of abortion have been equally ridiculous. The Health Administrator of the City of New York proudly announced that, after two years of legalization in the state, some of the "beneficial social consequences" of abortion were the decline of the infant mortality rate to an all-time low, and the first reduction in the number of out-of-wedlock births since the beginning of such records in 1954. This dubious manipulation of statistics by an equally dubious exercise of semantic wizardry came about as a result of the elimination of over 400,000 fetuses.

The issue may be defined simply. When does human life begin? Is the fetus at some stage a human being, and if so, at what stage? If the fetus is at no stage a human being then abortion should be perfectly acceptable. If at some point it is, then, equally, the destruction of the fetus after that point is murder.

The opposing views are illustrated by the late Dr. Alan F. Guttmacher, former president of Planned Parenthood Federation of America, and Dr. John R. Cavanagh, president of The Federation of Catholic Physician's Guilds. Dr. Guttmacher says: "My own speculations lead me to conclude that life begins only after the birth of a fetus sufficiently mature to survive outside its mother's body. Until then, there is only potential life, just as there is even before fertilization."[6]

Dr. Cavanagh takes a contrary view: "Life begins at the moment of conception. I know there is a difference of opinion, but I believe the child becomes a human being at the moment of fertilization."[7]

The biological argument for the belief that a human being exists at the time of fertilization is as follows. By any definition, the fertilized egg is living. The question is, is it a *human* life? All forms of life are defined by their characteristics, and determined by their genes. From the time of fertilization a human being has the genetic characteristics that will determine its makeup throughout its entire life. From the time of fertilization a life goes through a series of stages of development, both physical and psychological. Furthermore, the characteristics may come and go; not all are present at one time. For example, reproductive capacity is entirely absent before puberty, and declines with old age. Teeth grow and fall out and grow and, possibly, fall out again. Growth ceases after about the age of twenty. Since there is no stage at which one can say "this is a human life," and again say at other stages "this is not a human life," one is forced to conclude that a human life begins with conception.

I have difficulty accepting either the position that life (when I use the term "life" I mean human life) begins with fertilization or that it does not begin until birth. From a religious (moral) point of view, the difference between life and non-life is that a human being is composed of both a spiritual and physical component. This is at the base of, particularly, the Catholic definition of a human being. However, there are two biological facts that argue against the body–soul of an individual at the time of fertilization. The first of these is the formation of identical twins. Here is a case of the fertilization of an ovum by a sperm. If this is the beginning of life, it is an individual life. However, we know that that egg, fertilized as it is, can divide and result in two human beings.

The second biological fact involves the reverse of formation of twins. Dr. André E. Hellegers has reported six cases of human beings resulting from the combination of two fertilizations into a single individual. In this case, then, we have one individual disappearing.

My second difficulty with the beginning of life at fertilization belief relates to what could be described as a requirement for a natural respect for life. Studies on the subject of the elimination of fertilized eggs indicate that as many as 50 percent of all eggs fertilized are eliminated even before implantation. Writing in the British medical journal *Lancet,* Professor Charles Lowe and Dr. Colin Roberts, of the Welsh National School of Medicine, suggest that as many as 78 percent of pregnancies in young married women between the ages of 20 and 29 end in abortion, often without the woman having known she was pregnant. They believe that a woman's body may be able to detect naturally a malformed fetus, and abort it spontaneously.[8]

Considering the situation necessary for fertilization, it is not out of the realm of possibility to think of a woman regularly active sexually, and using no form of birth control, as producing a fertilized egg every month. When a woman ovulates, the egg is available for fertilization for at least 12 hours. The life of a sperm is at least two to three days, and there are reports of sperm remaining active for seven days. Consider, too, that during each sexual intercourse as many as 150 million to 250 million sperm are deposited. Any time the presence of the sperm overlaps with the presence of the egg, fertilization is possible.

In other animal species, fertilization at each ovulation is amost 100 percent predictable. While it is not practical to prove experimentally, the conditions exist in human beings to allow the same results. The difference is that the requirements for implantation in the human uterus are more sensitive and subject to more influencing forces than are those of lower animals. Therefore, there is the possibility that a fertilized egg could be eliminated every month with no evidence available that the event has occurred.

If these fertilized eggs are really human beings, I have great difficulty in accepting a theology that literally allows for the flushing down the drain of that many people. It is difficult, too, to accept the concept that the fertilized egg is human and must be protected, while recognizing that nature has no such respect.

On the other end of the scale, I cannot accept the time of birth as the time when humanness is conferred. This means, and the proponents of the position use the term, that *viability,* the ability to survive, is really the criterion for humanness. It is easy to picture the time in the not too distant future when medical techniques will progress to the point where more and more premature infants are being saved and where the time for development in the womb becomes less and less. With each new advance, then, the definition of humanness must change. As more research takes place it is likely that fetuses of 24 weeks of age or even less can be saved. In fact, the time is here now. On February 24, 1977, quadruplets were delivered twelve weeks prematurely in the Children's Hospital in San Diego. Only a very few years ago the chance for survival would have been zero. However, with the help of facilities available in the neonatal intensive care unit of the hospital, the infants, each of whom weighed a little over two pounds at birth, survived. Three and a half months after they were born three of the children had gone home and the fourth was doing well in the hospital. Physicians at the unit estimate that at present 50 percent of babies twelve weeks premature will live if they are given the treatment available. Thus, we are in a position of allowing for abortions of fetuses that could be saved, and are, therefore, human under the "viable" definition.

In attempts to define death, the detection of brain waves has been used as a criterion. Therefore, if the absence of brain waves defines death, why not use the presence of brain waves, appearing around the eighth week, to define life?

Physicians themselves distinguish the fetus from an abortus by implication when they call a fetus delivered after twenty weeks a "premature infant." Most physicians select the twenty-ninth week as the age of survivability.

Abortion is repugnant to me; I think it is wrong. This conviction is the result of considering the medical, moral and ethical points of view. I have said that I cannot accept either fertilization or birth as the criterion for the beginning of life. Neither can I find any convincing reason to accept any of the other times or events between these two—not quickening, not brain waves, nor hormonal activity, not survivability. I believe in the sacredness of human life. Therefore, in the absence of any convincing evidence for an alternative, I must choose the next biological phase after fertilization as the time when life begins, and when a fetus has legal rights and human status. I believe that this occurs at the time of implantation of the egg in the uterus, from one to two weeks after fertilization.

Death

Just as the abortion issue has stimulated a need to define the beginning of life, so the advent of new technologies in medicine has made apparent the need for a new definition of death.

The first indication of this need came with the introduction of organ transplants, when controversy arose over whether or not the organ donor was dead at the time of donating, for example, a heart. It has become possible to maintain some of the vital signs of life by mechanical means long after a patient would have been declared legally dead without the use of these supports. This has resulted in an increasing realization that the traditional definition of death no longer applied in some cases, and efforts have been made to refine it in the light of these new medical technologies.

The centuries-old criteria for death, easily observable, were when the individual stopped breathing and the heart stopped beating. However, the respiratory and circulatory functions are the very things that can now be maintained artificially by iron lungs and heart stimulators long after the individual's body has ceased to do these things normally.

With the advent of organ transplants there developed some public uneasiness relating to the treatment of the donor of the transplant.

It was suggested that, in the zeal to obtain a healthy, living organ for transplantation, proper consideration might not be given to the person from whom the organ was being removed and that, in fact, the organ might be removed while the donor was still alive. The difficulty was obvious. Many of the donors were selected from accident victims, victims who were so seriously injured they had no chance for recovery or who had, in fact, been pronounced dead but from whom an organ could be removed almost immediately while it was still viable. It was common practice to place these patients on respirators in order to keep bodily functions operating to assure a healthy organ. When the organ was removed, the respirator would be turned off. The question arose as to whether or not a patient on a respirator was really dead, in view of the old definition of death. The search for a new definition was accelerated by several court decisions that, because of the circumstances, developed new legal standards of death.

One of these cases will illustrate the problem. A man was brought into a hospital suffering from massive brain damage caused by an accident. Surgery was performed, but the prognosis for recovery was nil and the attending physician stated that death was imminent. The patient was placed on a respirator and was kept alive mechanically. An electroencephalogram showed no clinical evidence of viability and no evidence of cortical activity. In other words, the brain was dead. The respirator was turned off and, after about an hour, the heart and kidneys were removed and the heart transplanted into another patient.

A brother of the donor then sued for $100,000, claiming that the transplant team had hastened death by shutting off the mechanical means of support. A jury decided that the physicians were not guilty, that death had, in fact, occurred when the brain had died. It was a very unfair obligation to place on the jury, but they had no choice but to decide guilty or not guilty. The case, tried in Virginia, was hailed as "bringing the law up to date."

In an earlier case in 1958 the Supreme Court of Arkansas had apparently ruled against "brain death" as a criterion for death. A husband and wife were involved in an automobile accident, and the husband was pronounced dead at the scene. The wife was taken to a hospital where she remained in a coma caused by brain injury. She died seventeen days later. The case was brought to court when a petitioner claimed the two should have been judged to have died simultaneously, that, in fact, the moment of the wife's brain death was her moment of death. The court dismissed the case, quoting Black's Law Dictionary which defines death as "the cessation of life; the ceasing to exist; defined by physicians as a total stoppage of the circulation of the blood, and a cessation of the animal and vital functions consequent thereupon, such as respiration, pulsation, etc." The court also said, "Likewise, we take judicial notice that one breathing, though unconscious, is not dead."

But is one breathing with the aid of a respirator breathing or not breathing?

In March of 1978 in Chicago a fifteen-year-old youth received a beating that resulted in severe brain damage. The boy had stopped breathing once on the way to the hospital and again in the emergency room. Three days later, while on support systems, his brain had shown only a flat wave pattern, indicating complete and irreversible disfunction. The parents of the boy requested that the support systems be discontinued. However, because of the doubtful legal implications, the hospital authorities refused. The question was raised as to whether or not the assailant might plead that death was not caused by the beating but rather by the pulling of the plug on machines. The dilemma is reflected in the remarks of one of the hospital administrators who said: "If this had happened as a result of diving into a swimming pool it would be easy to disconnect the respirator, but with another party involved it would be a legal matter as to whether the party or the hospital was responsible for the death." After five days of painful delay, and only after assurances that the hospital would not be held responsible, the respirator was discontinued, and the boy's heart stopped beating fourteen minutes later.

The state of Illinois, several years before, had tried to anticipate such situations and had passed a brain death law. They had taken note of the several court cases involving the determination of the time of death when accident victims had been used as organ donors. The law passed referred specifically to a situation existing when organ transplants were involved, and made it legal for brain death to be used as the criterion in such cases. The parents of the injured youth had agreed that their son's organs could be transplanted, thus making a brain death decision technically legal. However, as one hospital administrator pointed out: "There was no way this could qualify as a transplant case. Since the Medical Examiner was certain to perform an autopsy, the organs would have been useless for transplant."

The Cook County Medical Examiner did perform an autopsy and concluded that the youth died of the effects of an injury. The State Attorney's office has said: "The action was proper and in accordance with the law in Illinois." The Anatomical Gift Act allows for brain death as a criterion. Other deaths are determined by the usual heart–lung standards. The alleged assailant has pleaded not guilty to the murder charge.

A further complication into a definition of brain death has been brought to light as the result of an occurrence in Zurich, Switzerland. Dr. Urs Peter Haemmerli, Chief of Medicine at the Triemli City Hospital, was accused of murdering an unnamed number of elderly patients at the hospital. The patients, as described by Haemmerli, were terminally ill, unconscious, and never going to regain consciousness. They were still breathing, and were being fed intravenously. At some point, the decision, concurred in by the entire medical staff, was made that the patients could never recover, and intravenous feeding was discontinued. Saltwater was substituted for the nutritious intravenous fluid in order to prevent dehydration. Death, in fact death by starvation, occurs in a couple of weeks. Haemmerli calls this practice sound and humane, and says it is painless.

Haemmerli's procedure, to be justified, requires a further definition of brain death. At present proposals deal with total brain death. However, the patients under discussion do have a type of brain death, but have preserved their spontaneous respiration. Haemmerli believes it is just as artificial to keep these people alive by artificial feeding as it would be to keep them alive by means of a respirator.

The desire to redefine death has spawned a whole series of discussions as to what death is, how it differs from life, whether it is a single, instantaneous phenomenon or a series of events taking place over a period of time. The single, last convulsive gasp, the final death struggle, has been the symbol of death—the moment of death—dramatized in literature. But there are those who argue that death is a process that takes place, not in an instant, but over a period of years. After an individual reaches the age of about twenty there is a continual decline. As he reaches later years whole organs deteriorate, until at very old age there is no resemblance to the vigorous bodily structure, or in most cases mental ability, of his youth. It is this deterioration that some people now call dying, and it is this philosophy that allows some to think that death could and should be accelerated by external means.

But is this really dying? I prefer to make a clean distinction between dying and aging. The process I have described is aging, and the individual is not dead, regardless of the deterioration of individual organs, until the whole organism is dead. Even the term "organism" causes difficulty, particularly if we associate it with the word "human." A human organism could be thought dead when those things that make it human die, that is, the death of higher functions such as the central nervous system and the cerebral cortex.

If there is no moment of death, then there must be a gradational time, differing possibly for each individual, when a person can be declared dead, obviously involving a subjective decision on the part of the physician.

If we assume that there is no "moment of death," that a person dies progressively, it is easy to reason that the value of that life decreases as death progresses. The next step becomes even easier. The less value a life has, the more justification there is for ending it, and positive euthanasia becomes only a natural factor in facilitating the process of dying.

Regardless of whether or not death is considered a long process, it should be possible to distinguish between living and non-living things. If there is a difference, then, there should be a time at which one passes from a living thing to a dead thing. This time is not when a limb is lost or an appendix is removed or when cells deteriorate. It is when an individual as a whole ceases to function. Possibly the last deathbed gasp is, indeed, the visible indication of the moment of death.

What has caused the confusion now is that, by technical means, we have obscured the indicators of death that have been used in the past. It does not mean that they are not still there. It just means that we now do not allow them to become obvious. For this reason, then, it is necessary to search for other signs that would indicate the same thing—the ceasing to function of the entire organism.

The first legislative attempt to define the time of death by statute occurred in Kansas in 1970. The law, stimulated primarily by the desire to have some legal basis for facilitating transplant therapy, has been criticized for a number of shortcomings. The law reads:

> A person will be considered medically and legally dead, if in the opinion of a physician, based on ordinary standards of medical practice, there is the absence of spontaneous respiratory and cardiac function and, because of the disease or condition which caused, directly or indirectly, these functions to cease, or because of the passage of time since these functions ceased, attempts at resuscitation are considered hopeless; and, in this event, death will have occurred at the time these functions ceased; or
>
> A person will be considered medically and legally dead if, in the opinion of a physician, based on ordinary standards of medical practice, there is the absence of spontaneous brain function; and if based on ordinary standards of medical practice, during reasonable attempts to either maintain or restore spontaneous circulatory or respiratory function in the absence of aforesaid brain function, it appears that further attempts at resuscitation or supportive maintenance will not succeed, death will have occurred at the time when these conditions first coincide. Death is to be pronounced before artificial means of supporting respiratory and circulatory functions are terminated and before any vital organ is removed for purpose of transplantation.
>
> These alternative definitions of death are to be utilized for all purposes in this state, including the trials of civil and criminal cases, any laws to the contrary notwithstanding.[9]

The law appears to be deficient in several respects. In the first place there are two alternate definitions of death. People cannot help but be confused about the possibility of two different kinds of death. Such a concept certainly does not do much to build up confidence in medicine when the conclusion might be drawn that, in the case of two people who have reached the same stage of dying, one might be pronounced dead and the other would not be. The objective of the second paragraph is probably to give the physician the legal right to turn off supportive treatment such as a resuscitator. However, presenting the situation as a second alternative is both clumsy and unnecessary.

The second shortcoming of the bill lies in its potential lack of protection both for the patient and the physician. The bill allows all of the decisions to be made by "a physician." Almost routinely, professional bodies who have considered this problem have required that at least two physicians should be involved with this pronouncement of death. In the case of decisions relating to transplants, the physicians making the decision have been specified to have no involvement with the subsequent transplant operation. In other words, some process must be followed that will assure the total objectivity of the decision, and that a resulting action such as the removal of an organ has absolutely no effect on this decision. In the proposed bill there is no effort to separate the decisions affecting the donor from decisions affecting the recipient of an organ.

Of greater use in the creation of legislation on death has been a report from the Ad Hoc Committee of the Harvard Medical School to Examine the Definition of Brain Death.[10] The Committee outlined four criteria for determining brain death, and stated, with great detail and specificity, what tests were to be run and what standards were to be met before declaring a patient dead. The four areas of test were (1) unreceptivity and unresponsivity to externally applied stimuli and inner need; (2) no movement or breathing for at least an hour; (3) no reflexes; and (4) a flat electroencephalogram. It was specified that all of the above tests be repeated at least twenty-four hours after the original tests were made.

The above criteria, obviously, do not replace the traditional criteria of death, applicable in most cases. They do apply to those cases where, for example, a person is on a respirator and spontaneous disappearance of natural functions is obscured.

Both the Harvard report and the Kansas law suggest that when death has been pronounced, only then may the respirator be turned off. This is obviously a protection for the physician. However, while the Kansas law states that a physician makes the decision, the Harvard report says, "The decision to do this [turn off the respirator] and the responsibility for it are to be taken by the physicians who have been directly involved in the case. It is unsound and undesirable to force the family to make the decision." They further suggest that the decision to declare the person dead be made by physicians not involved in any later effort to transplant organs or tissue from the deceased individual.

Finally, there is the question of whether or not legislation is needed to help solve the problem of the definition of death. I believe a new definition is necessary. But *who* will define death?

The courts could make this decision through case law. However, this would be a most unsatisfactory solution. In the first place the eventual definition would probably have been built over a period of time as the result of several decisions involving different points of law. Second, the degree of medical input would not be of the same quality as that which would result from expert medical discussion in a form other than a court of law. Similarly, a court of law would not be a proper atmosphere in which to have opinions from the general public.

The final objection to this method of approach is the natural reluctance of the courts to break any new ground in the field of medicine or medical ethics.

The most important, certainly the most publicized court case concerning the definition of death is that of Karen Quinlan. Following a party in mid-April, 1975, Karen fell into a coma. She was placed on a respirator to maintain breathing. Five months later, after no improvement, and after medical opinion concluded that she was irreversibly brain damaged and would never recover, her parents requested that the support system be discontinued. Although the Quinlans gave their consent, the attending physician and the hospital refused to remove the respirator.

The Quinlans then asked the court to declare Karen incompetent and to appoint her father as guardian. Not only did the court not grant the right to remove the life sustaining procedures, but it appointed a stranger to the family as Karen's guardian.

The case was appealed to the New Jersey Supreme Court, which reversed the decision of the lower court. Karen's father was appointed her guardian, and permission was given to withdraw the support system if the attending physicians agreed there was no hope for recovery and if an ethics committee, established by the hospital, concurred. It seemed that after more than a year of anguish, the ordeal of the Quinlans would finally be over. But even in view of the court decision, the attending physician and the hospital administrators refused to discontinue treatment. Instead, over a period of only a week, Karen was weaned from the support system. Three years later she is still alive.

It should be remembered that this case did not represent one that involved a definition of brain death. Karen did not meet the criterion of a flat brain wave. The case is an example of a tortuous procedure followed before recognizing the right of a patient or the patient's guardian to refuse treatment.

If courtroom procedures are too restrictive, as I think they are, to permit the presentation of the many viewpoints that must be considered in arriving at a definition of death, it seems equally undesirable to have death defined, possibly in fifty different ways, by laws promulgated by each state. It is apparent that judgments relating to the time of death depend on the definition of death. Legal determinations of the order of the survival of heirs then, for example, could vary in different areas in the country if the definition of death is different. Insurance claims would vary also, and what might be considered malpractice in one state would be perfectly legal in another. The alternative to having the eventually accepted definition evolve as a result of a series of court cases is to have a definition develop at the Federal level to be applicable in all states. It is here that public discussion and debate can be most open, and where every factor involved can be subjected to public scrutiny and judgment.

I have emphasized the desirability of having the public take part in discussions, because the problems are not just medical. They are moral, ethical, religious, and legal as well. However, the decision that will be made as to the time of death, regardless of the criteria set, will be strictly that of the physician. He will interpret medical data and will decide whether or not it meets the criteria laid down for death.

A third method of establishing a definition of death would be to have no statutory change in the law, but to have standards determined by and agreed to by the medical profession. It would be the responsibility of the physician in charge of any patient to operate under the proposed guidelines, in consultation with other physicians.

This third alternative, considering the interpretation of these criteria to be solely a medical responsibility, would probably be the most desirable if it could be made to work. However, there are enough examples available already to show that medical consensus is not sufficient to protect either the patient or, more recently, the physician. Without the force of law behind the decision, every physician is liable to be sued for malpractice.

Legislatures, as evidenced by at least seventeen states, are now willing to face the problem of defining death. Alaska, Kansas, Maryland, New Mexico, Oregon, and Virginia allow for either brain death or the absence of spontaneous respiration and circulation, thus encountering the problem of "two kinds of death." Iowa, Louisiana, Michigan, Montana, and West Virginia preserve the traditional respiratory–circulatory death as the primary means of determination, but extend the definition to include brain death if artificial means of support obscure the fact that these functions have ceased. California, Georgia, Idaho, Illinois, Oklahoma, and Tennessee have defined death as "irreversible cessation of total brain function" without regard to the traditional criteria.

I believe that the acceptance of brain death as a criterion is now generally held. There is no real disagreement between those who accept brain death and those who retain heart–lung functions as standards except tradition.

However, another area of discussion is now arising between those who want the requirement to include the death of the whole brain and those who say that the destruction of only a part of the brain, that controlling the higher mental functions, should be sufficient.

The lower brain controls blood pressure, respiration, reflexes, etc. The higher brain—the cerebrum—controls those things that relate to "consciousness"—memory, ability to reason, rational thought—in other words, those things that distinguish human beings from other animals.

It is now being suggested that if only the functioning of the higher brain is destroyed death has occurred. The rationalization of this position is that there is a distinction between the biological *organism* and the biological *person,* that when those things that distinguish a person from an organism are eliminated, death has occurred.

My own position would be that there are too many hazards involved in the definition of personhood to allow such a determination to define death. I am in total accord with the criterion of total brain death. It is definable and measurable. To include a requirement for personhood, a concept on which there is no, and possibly can be no, agreement, is to allow for varying judgments on a matter that should not be susceptible to personal, even whimsical, interpretation.

What is death? It is the end of living. It is just as much a part of the total process of life as is birth. Just as birth is a progressive development from conception through fetal growth, so death is a series of events. Clinical death occurs when respiration and heart rate cease. Brain death follows clinical death unless resuscitation is instituted. Following clinical death and brain death, cellular death takes place. The cells die slowly. Cells from some parts of the body can be grown in tissue culture, indicating life, when taken hours or even days after clinical death.

The brain too dies in stages. First the cortex dies, then the midbrain, then the brain stem. Higher levels of the brain may cease to function without damage to vital centers in the lower levels of the nervous system. The heart may beat and respiration continue even though there is permanent loss of consciousness. When all the components of the brain die all bodily life disappears.

From the above it can be concluded that what we now call clinical death can be postponed by artificial means. But there is no method by which the brain, once it has ceased to function, can be kept alive artificially. There are no pacemakers, resuscitators, or oxygenators for the brain. Even though circulation and respiration can be maintained artificially, if the brain is dead the individual can never be a functioning person. Thus the argument for using brain death as the criterion for human death, and the argument for the futility of using artificial sustainers of bodily life after brain death. I suppose it is natural to try to postpone death as long as possible. Thomas Browne said, "The long habit of living indisposeth us to dying." About fifty million people die each year. With a handful of exceptions everyone now living will surely be dead in the next one hundred years. With such inevitability, why the fear of death, the intense desire to postpone it, the belief that, whatever the circumstances of living, living is better than dying? If dying is normal, natural and inevitable, would it not be better to cooperate with the process of dying, rather than to obstruct it?

I have no difficulty in accepting completely the teaching of Christian moralists that extraordinary means need not be taken to prolong the life of an individual. This teaching received considerable publicity when, in November of 1957, Pope Pius XII spoke to a group of physicians in Rome, saying: "But normally one is held to use only ordinary means—according to circumstances of persons, places, times, and culture—that is to say, means that do not involve any grave burden for oneself or another."[11] And further and specifically, he continued:

> The rights and duties of the doctor are correlative to those of the patient. The doctor, in fact, has no separate or independent right where the patient is concerned. In general, he can take action only if the patient, explicitly or implicitly, directly or indirectly, gives him permission. The technique of resuscitation which concerns us here does not contain anything immoral in itself. Therefore, the patient, if he were capable of making a personal deci-

> sion, could lawfully use it and, consequently, give the doctor permission to use it. On the other hand since these forms of treatment go beyond the ordinary means to which one is bound, it cannot be held that there is an obligation to use them nor, consequently, that one is bound to give the doctor permission to use them.[12]

Sir William Osler, the great medical pioneer, was one of the strongest proponents of the idea that death was a normal, even desirable process, and should be accepted as such. He objected, in particular, to the phrase "agony of death," stating that there was no such thing.

There is a considerable body of evidence to support this position. Many people are living today who have gone through the process of dying. Those resuscitated from cardiac arrest have been technically dead. Of those interviewed practically none recall anything approaching agony. The predominant feeling was one of peace, even detachment.

Unusual as it seems, we should give up the idea that death is a defeat, that it is a catastrophe or, most of all, that it is avoidable. As long as we believe that we die as a result of being sick, the idea of preventing sickness will be confused with preventing death. But even if we could conquer everything that we call disease we would still die.

Why, then, the furious energy going into preserving life, even a life that has long since outlasted its usefulness to contribute to society or even to savor the self-satisfaction that comes from accomplishment? I believe much of the impetus to use extraordinary means comes, not just from a desire to preserve a human life, but to prevent defeat. About 75 percent of all U.S. residents die in a hospital. I am convinced that those charged with hospital care—doctors, nurses, technicians—look upon preventing death as a challenge, a challenge to battle, and a battle to be won under any circumstances. The doctor is never commended when a patient dies. Instead he or she has "lost a patient." We speak of the "victory of death" when, in fact, we should not even think of a conflict.

Hospital staff, for many reasons other than humanitarian, do not want patients to die. It is, after all, a great inconvenience. Legal requirements of record keeping must be met, treatment or non-treatment must be justified. Thus, it is always better to overtreat than to appear to be undertreating. The result is the tube up the nose and down the throat, the implanted intravenous tube, the lights that blink, the buzzers that buzz, and the bells that ring. It is sometimes difficult, observing a patient under such circumstances, to decide whether the instruments are responding to the patient's reactions, or whether the patient is just another organism controlled by the technological gadgets employed for another laboratory experiment.

Even after death the game goes on. We pretend the dead are still alive. Thus the elaborate preparation of the body for display. "How natural he looks!" But how can he look natural? How often when alive was he seen lying stiff in a satin-lined box, cosmetized, hands crossed, surrounded by flowers and, possibly, lighted candles?

If they are not to "battle death," what should be the duties of physicians? Only to use ordinary means of treating the patient and, if requested, to use extraordinary means if available. On the other hand, it is the clear right of the patient to refuse extraordinary treatment or, if the patient is not capable of making such a decision, a legal guardian or other representative has the same right. The doctor is relieved of any responsibility at this point.

The objective should not be just to extend life under any circumstances. Rather we should try not to eliminate death, which is impossible, but to eliminate untimely death. Centuries ago Seneca said that, "The wise man lives as long as he should, not as long as he can." Death, then, is bad for the individual only if it is untimely. For society, death is a necessary benefit. Death is nature's way of rejuvenating the population, of eliminating those who have fulfilled their purpose in the world, of making room for those who can take up where past generations, exhausted in ideas, have left off. The only difficulty with this concept is the ability to decide what is untimely.

Attempts have been made to formalize the already existing right of patients to reject treatment even if it means death, by developing laws that would allow the patients, while still physically and mentally able to do so, to write into a "living will" their desire that no extraordinary means be used to prolong life.

Late in 1976 California passed a Natural Death Act. Essentially, the law allows the patient to direct the doctor to eliminate life support care when natural death is imminent. It specifies that the patient "has voluntarily and in sound mind evidenced a desire that such procedures be withheld or withdrawn." While such bills have as their ostensible purpose the protection of the individual to die with dignity, the major impetus to such legislation comes from the desire of the medical community to clear up the uncertain legal status of the physician in cases where life support systems are either withdrawn or else not used in the first place. The proposed bills do release all medical personnel, as well as the hospitals and other health facilities, from civil liability and criminal prosecution. The California bill also states that when the patient orders the cessation of the life support systems he or she shall not have committed suicide.

Since the California law was passed, legislation involved with the "death with dignity" issue has been introduced into at least 39 other state legislatures, and has been passed by seven of them. There has been no attempt to standardize these policies. Some of them would even legalize active killing, or positive euthanasia, to use the more genteel term.

In the effort to allow a person to die with dignity, it is easy to overlook the rights of the patient. In any legislation three important factors must be borne in mind. First, the person should be allowed to die, not be actively killed. Second, the patient must have chosen to die rather than be treated with extraordinary means. Third, if the patient is incompetent, it must be clearly understood who can make the life or death decision in his or her name.

One of the things thought to have eliminated some of the legal problems involved was the recognition for the first time of the legality of the so called "living will." The living will is a document that directs the physician not to use extraordinary means of sustaining life when death is imminent, and gives the physician permission to terminate such procedures if they have already been instituted. Hundreds of thousands of people in the United States have signed such wills.

This sounds like a step forward, but I don't believe it is.

By implication, it is based on a misconception of the patient–physician relationship. Physicians have no rights at present over patients, nor did they ever have. The patient always has the right to accept or reject treatment prescribed. This position is supported by law. Without the consent of the patient, physicians may be liable to criminal action for assault and battery.

If the rights of the patient were recognized, it would not be necessary to have additional laws, nor would it be necessary to recognize the living will as a legal document. If the patient is unconscious and unable to communicate, it is the right of the family to make a decision based on the presumed will of the patient.

In the light of these presumed rights of the patient, there is, then, one undesirable consequence of the legalization of the living will. Such a will is reassuring to the family and probably to the physician when it is in existence. However, there is a danger that, in the absence of a will, the presumption would be that the patient would be overtreated, that, as seems to be prevailing opinion now, the decision making power would rest in the hands of the attending physician who would take every precaution for self-protection. Since there will probably always be many more people dying without such wills than with them, the erroneous idea of the right of the physician over the patient would be strengthened further by the implication in the law.

There is no obligation to preserve a patient's life merely because it is medically possible to do so. The assumption that extraordinary means need not be used is becoming more generally accepted. However, there is some misconception about what is meant by ordinary and extraordinary means. I am basing the assumption on a theological, rather than medical, definition of "extraordinary."

In medical terms extraordinary means would be a treatment not in general use, possibly an experimental procedure or one involving grave risk. In these days the use of ventilators and resuscitators, dialysis units, even some organ transplants would be considered ordinary means. The theological definition is based on the needs and reactions of the patient, rather than on technology. In the address referred to previously, Pius XII said "But normally one is held to use only ordinary means—according to circumstances of person, place, times, and culture—that is to say, means that do not involve any grave burden for oneself or another."[13] Thus, although some treatments might be considered ordinary medically, they would be extraordinary from a theological point of view. For example, treatment that would involve a very heavy financial burden on the patient or his family would be extraordinary. So would a treatment that caused excessive pain, or one that might render life after treatment unbearable.

In all cases of extraordinary means the patient, acting independently or through his or her family, is free to reject it, and the physician has no right or even responsibility to insist upon it.

Because of the ambiguous terminology of "usual and unusual," the terms "reasonable and unreasonable" are becoming more common as expressions of what is required or not required for treatment. I think these terms express more accurately what is desired. A treatment that is "usual" in medicine might be judged "unreasonable" in some cases. An "unusual" medical treatment might be "reasonable." The terms reasonable and unreasonable also do one other positive thing—they make it evident that the judgment to be used in making a decision, either by the patient or the guardian of the patient, is based on reasonableness, not on usualness. The emphasis, then, is on "reasonableness for the individual," not usual for the individual as one of a category.

Changing a word doesn't necessarily eliminate the problem. Whether treatment is usual or reasonable, or unusual or unreasonable, someone must decide what the words really mean.

In Massachusetts a two-year-old boy was diagnosed as having acute lymphocytic leukemia, a form of cancer. He was started on the orthodox treatment for the disease—frequent pills and injections. The side effects of such treatment are usually severe—frequent vomiting, loss of hair, moon-face caused by steroids—but it is the only effective treatment available. After a month of treatment the disease was in remission. Another month and a half of weekly spinal injections of a drug was instituted. This was accompanied by injections of an enzyme and later anti-cancer pills.

Without telling the physician, the parents of the boy decided to discontinue the medicine that was to have been administered at home. Four months later the physician learned that the leukemia had returned. After trying, without success, to convince the parents that drugs were the only chance to keep the boy alive, hospital lawyers instituted proceedings to force the parents to agree to treatment. In late March of 1978 a District Court judge ruled that the parents did, indeed, have the right to discontinue their son's treatment. The hospital immediately appealed, and in April a Superior Court judge ordered the treatment resumed.

Who decides what is reasonable, the court or the parents? Physicians say that without orthodox treatment the boy will surely die. The parents say the treatment is "poisonous," that the child can be cured by diet and other treatments.

The courts, in the past, have ruled in favor of compulsory medication as, for example, when the Supreme Court ruled that forced vaccination was legal, and also ordered blood transfusions for children of Jehovah's Witnesses.

The question of euthanasia has, inevitably, been raised in these considerations. Even here there is semantic confusion. The Greek word "euthanasia" originally meant a "good death," and this is still the preferred dictionary definition. However, almost universally, euthanasia has come to mean an "accelerated death by painless means," and there are now two different kinds of euthanasia, active and passive. Active euthanasia is causing a death to happen, or accelerating it. Passive euthanasia is *allowing* a death to happen, even though means might be available to prevent it.

With the acceptance of the moral and legal correctness of passive euthanasia, a term I dislike very much but which I shall use because it is expressive, came fears that the next step would be active euthanasia—infanticide and the elimination of the old and the physically and mentally disabled.

Such fears are not groundless. In 1973, 86 percent of the members of the Association of Professors of Medicine favored passive euthanasia, while 18 percent favored active euthanasia. In 1974 a Gallup Poll indicated that 53 percent of the American people believed a physician should be allowed, by law, to end painlessly the life of a person with an incurable disease. In 1950 only 36 percent had answered the same question affirmatively.

In 1977 alone at least seven bills that would legalize euthanasia were introduced in this country's state legislatures. From 1970 to 1972 forty-three severely defective newborns were deliberately not treated for other diseases and allowed to die at the Yale–New Haven Hospital. In 1971 a Down's Syndrome infant was allowed to starve to death at Johns Hopkins University Hospital. It required fifteen days.

Dr. William McNath, medical superintendent at Neasden Hospital in northwest London, instructed his staff on which patients should be resuscitated and which allowed to die in the event of heart failure. Persons designated "not to be resuscitated" included those over 65, and those suffering from malignant diseases, chronic chest diseases, and chronic kidney diseases. No other criteria were specified. In response, the Ministry of Health issued a statement ruling that no patient should be excluded from consideration for resuscitation by reason of age or diagnostic classification alone.

Earlier I used the term "cooperate with death." Possibly a better expression would be to say we should accommodate to death. As death approaches, life and death coexist. Whether we think about cooperation or accommodation the important thing is that death not be speeded up by any means nor life prolonged by extraordinary means. The difference between causing death by positive action such as injecting air or an overdose of drugs, and allowing death to occur is admittedly a fine distinction. Yet there is a distinction and I think a clear one. While it may be said that injection on the one hand and turning off a resuscitator on the other are both positive actions with the result being the same—the death of the patient—there is a difference. Each is an extraordinary means. Without the use of the injection as a means the patient would stay alive. Turning off the resuscitator as a means results in death. However, death as a result of an injection would not be normal in the way we are using the word, while death without the resuscitator would be normal.

If the result is death in both cases the argument has been made that the distinction between positive and negative euthanasia is simply one of semantics. If the result is the same why the objection to positive euthanasia? Totally aside from the moral and legal issue of whether or not positive euthanasia is murder there are two reasons for resisting this practice. The first involves the remote possibility that a diagnosis or prognosis might be wrong. In the case of positive killing, death would result. In the case of withholding treatment the patient would have a chance to survive. The most celebrated example of this, mentioned earlier, is the situation of Karen Quinlan, who is still alive three years after support systems were removed.

The longest known time for a person to have been in a coma is 37 years. In August of 1941 six-year-old Elaine Esposito, then living in Chicago, was anesthetized for an appendectomy. She went into a coma and never recovered consciousness. Through the years she was cared for at home by her mother and father. Elaine's father died in January of 1978, leaving to Elaine's mother full responsibility for her daily care, including the necessary feeding through a tube in the throat. Elaine died on November 28, 1978, in Florida, where the family had been living for the last 18 years.

Humanists may argue that this really supports the argument for active euthanasia—that compassion dictates that both the patient and survivors would be better off if the desired death had occurred. Regardless of the results, the moral position requires that the patient be allowed to live or die, and not that she be made to die.

By far the most convincing argument for me, however, is the possibility of the dehumanization of those practicing euthanasia and, eventually, the dehumanization of the society that permits and condones it, with the result that there could be abuse of the practice. The interest of the patient always comes first. He or she is morally and legally the decision maker, with the right to refuse treatment. It would be easy, if positive euthanasia became accepted practice, to put too much emphasis on the interest of the survivors responsible for the patient. It would be easy to develop an atmosphere in which anyone not able to contribute to society would be eliminated. It would be easy to rationalize this as being in the interest of the patient, when, in reality, it was only for the convenience of guardians or relatives.

In *Julius Caesar* there is an exchange between Casca and Brutus that illustrates the point very well:

> *Casca:* Why, he that cuts off twenty years of life
> Cuts off so many years of fearing death.
> *Brutus:* Grant that, and then is death a benefit;
> So are we Caesar's friends, that have abridg'd
> His time of fearing death.[14]

It is my conviction that positive euthanasia is never justified. I think omitting or ceasing treatment is justified in cases where medical opinion concludes there is no hope of recovery. The definition of death should be based on brain death, and not on clinical death. Finally, the patient must have the right, if he or she is mentally competent, to accept or reject unusual treatment, even though it may be lifesaving.

3 . . . *wherein many and diverse genetic defects are explained, and they exist in good and bad and obvious and hidden forms*

Were it not for the fact that a tiny bit of chemical, so small as to be invisible under even the most powerful light microscope, had malfunctioned in a single human body the American states might still be colonies of England. Based on the recorded fact that the urine of George III of England was dark red in color, medical historians are in general agreement that he suffered from porphyria, a genetic disease whose symptoms are caused by the massive buildup in the body of porphyrins, chemicals that are intermediates in the body's synthesis of hemoglobin. The King exhibited the usual symptoms: convulsions, frequent hallucinations, unstable and manic-depressive behavior. With the knowledge that we possess today he could undoubtedly have been treated in such a way that he would have been in a position to make more rational decisions. Without the oppressive actions taken by him while in an irrational state, would the colonists have been forced into rebellion?

All genetic diseases do not give occasion for such dramatic speculation. But all genetic diseases, all genetic changes, take place as the result of a change in just such a minute portion of the organism as that which influenced George III and, perhaps, changed the course of history.

It is only now that we are beginning to recognize how these tiny bits of chemicals, chromosomes and genes, control the evolution of living things, in some cases causing the appearances of new species, in others profoundly influencing the physical and mental abilities of individuals.

It has been 120 years since Charles Darwin concluded his book

Origin of the Species by Means of Natural Selection with these words:

> Thus, from the war of nature, from famine and death, the most exalted object which we are capable of conceiving, namely, the production of the higher animals, directly follows. There is grandeur in this view of life, with its several powers, having been originally breathed by the Creator into a few forms or into one; and that, whilst this planet has gone cycling on according to the fixed law of gravity, from so simple a beginning endless forms most beautiful and most wonderful have been, and are being, evolved.

Darwin built his theory on the information available to him through the study of previous theories, and through direct observation. In the last forty years information has been developed, not just as the result of observation, but of experimentation, that was probably not even dreamed of by Darwin. Now we have a basis for understanding not just that evolution has come about, but how it comes about. And even more awe-inspiring, we now have methods of causing it to come about.

The fact that we now understand more of the mechanisms of evolution, of genetics, does not in any way detract from the feeling of appreciation for the grandeur of this master plan of life, with its combination of incredibly complex and beautifully simple elements.

It has taken nature between four and five billion years to produce the variety of life existing on our earth today. As fellow occupants of our earth we have, as codependents, about one million animal species and a quarter of a million plant species. Since nature is conservative about change, a new species does not come into existence and persist through time without considerable testing. About 500 million other animal species came into being through the ages but failed nature's test, and did not endure.

We humans are latecomers to the biological scene. We have been present on earth much too short a time on the scale of evolution for any significant change to have taken place in our species. The genetic systems common to us now are the same systems that were present at the beginning of civilization. The same biological brain that resulted in the philosophy of Socrates, Plato, and Aristotle, produced the mathematical miracles of Ptolemy, and resulted in the genius of

Watson and Crick, was probably in existence many thousands of years ago.

The exquisitely sensitive and incredibly complicated process involved in the duplication of members of the species from generation to generation, or in the change within a species, sometimes involves a single event, and other times a thousand steps and millions or billions of years. Some species remain unchanged for eons. The cockroach has been around for at least 250 million years. Fossil remains of prokaryotes, organisms whose cells contain no nuclei, the simplest form of life on earth, have been found dating back 3.4 billion years. Yet, five hundred species have disappeared for every species that now exists.

Through all these billions of years, it is nature that has controlled evolution, but the situation may be changing. We now know how genetic traits are transmitted and this puts into our hands a tremendous power to be used for good or for evil. It allows us to consider genetic manipulation. It allows us to develop screening procedures for genetic defects. More than that, it allows us to predict the results of genetic interactions, and thus to act to avoid the undesirable and to encourage the beneficial.

The discovery that deoxyribonucleic acid (DNA) is the chemical basis of heredity is probably the greatest discovery in the history of biology. The Austrian monk, Gregor Mendel, first stated a theory proposing a "unit of heredity" in 1866. It was not until after the turn of the century that scientists could study the individual cell, breaking it down into its nucleus and other components, and isolating acids, called nucleic acids, from the nucleus. Not until 1940 was it proposed that these acids, DNA and RNA (ribonucleic acid), might be the transfer agents of genetic characteristics.

It has been about a quarter of a century (in 1953) since the brilliant work of Watson and Crick determined the geometry and chemical composition of the DNA molecule. DNA consists of two long strands of molecules, twisted around each other in much the form of an elongated double pretzel, the famous double helix.

Genes are the basic units by which individual characteristics or traits are passed on from generation to generation. They consist of very specific interlocking strands of DNA. Each gene is responsible for the determination of a particular protein. The proteins, equally specific, then become part of whatever trait or characteristic the gene has ordered. Genes, in turn, are distributed along the length of the chromosome, much as beads are strung on a chain. The DNA molecules, or genes, clumped into chromosomes occur in the nucleus of every one of the trillions of living cells that make up the human body. The way in which these chemicals are arranged determines whether the resulting organism is to be a fish or a bird or a human. They determine the characteristics of the organism—sex, size, eye color, etc.—for the rest of its life. They are present at the first fertilization of the egg. With each cell division they will be duplicated again exactly unless some intervention occurs. It has been estimated that there are 60 trillion cells in the adult human body. The genetic "code" of an organism is present in exactly the same form from conception to death. Thus, DNA is truly the master blueprint from which an organism is built.

Based on estimates of the number of genes in lower species and then extrapolating to humans, and based on the amount of DNA present in humans, and other mutational limitations, it is reasonable to assume that there are roughly 50,000 genes present in the human body.

However, it has been only since 1956 that the correct number of chromosomes in each individual has been known. Each male and each female has twenty-three pairs of chromosomes. Twenty-two of these pairs are non-sexed chromosomes, called autosomes, and the other pair are termed sex chromosomes. In the female the sex chromosomes are identical, and identified as X. In the male the sex chromosomes are different, and consist of an X and a Y chromosome.

The inherited characteristics of an individual are determined by the combination of chromosomes contributed by the mother and the father. Each sperm and each egg has twenty-three unpaired chromosomes. At the moment of fertilization the chromosomes from the parents are paired, determining the characteristics of the individual for the rest of his or her life. For each characteristic the new person has received an approximately equal contribution in the form of an individual gene from the father and from the mother. Thus, each cell will contain forty-six chromosomes, and each cell will contain two sets of each gene.

Genetic defects, or genetic diseases, are caused when there is an abnormal arrangement in the DNA sequence of even one of the thousands of genes in the ovum or sperm. The difference between normal and abnormal can be very small, but incredibly specific. For example, sickle cell disease results from the variation of just one amino acid out of the 574 present in the hemoglobin of the red blood cells.

Changes or mutations are quite common in nature. They may occur spontaneously, or they may be stimulated by cosmic or X-rays, chemicals, or even stresses such as hard blows or physical irritations. More often than not, because most are lethal, the changes will pass completely unobserved. However, sometimes they do persist.

Regardless of how the change comes about, there is a chance that the same defective gene will be passed on to the progeny. The probability rate of that chance depends on the type of gene involved. With our knowledge of genetics the statistical likelihood of offspring being affected can be predicted for a large number of diseases.

I have already mentioned that, in the male and female, twenty-two of the twenty-three pairs of chromosomes are identical, and are called autosomes. The twenty-third pair differ in the male and female: these are the so-called sex chromosomes. In addition, genes are classed as dominant or recessive, depending on whether or not the trait conditioned by a single gene can be neutralized or expressed regardless of the contribution by its companion gene in the pair. For example, if only one parent possesses an abnormal gene for a particular trait, the offspring would inherit one abnormal gene and one normal gene to make up a gene pair. If the gene in question is recessive, the normal gene in the pair would mask the effect of the abnormal gene. However, if the gene is dominant, its action is not masked, and the abnormal trait is evident. If both parents possess the abnormal recessive gene, the offspring would then inherit a double dose of the gene, and would be susceptible to the disease. Diseases can, therefore, be classed as autosomal dominant or recessive, or sex-linked dominant or recessive.

An individual who has one defective gene and one normal gene is said to be "heterozygous." An individual in whom both genes are either normal or defective is called "homozygous."

With this information, then, it is possible to establish the statistical probability of a trait being transmitted to the next generation. Prospective parents, each of whom carry the same autosomal recessive gene have a one-in-four chance of having an offspring with the disease. This is a statement that is frequently misunderstood. It does not mean that if a couple has four children one would be predicted to be abnormal and the other three normal. It is a statistical result of the analysis of the possible ways in which two gene pairs, each of which contains one normal and one abnormal gene, could combine. The results can be two abnormal genes, two normal genes, or two pairs containing one normal and one abnormal. Since we are talking about a recessive gene, another way of stating the prediction would be to say that there is a 25 percent chance of a resulting child inheriting neither of the defective genes and therefore being normal; a 25 pecent chance of inheriting both of the defective genes and therefore being afflicted with the disease; and a 50 pecent chance of a child inheriting one normal and one abnormal gene and, therefore, being disease-free but a carrier of the defective gene. This does not mean that in a family of four children one will be normal, one will be abnormal, and two will be carriers. The analysis refers to each individual pregnancy, regardless of what has gone before. A family could have four normal children without having any who have inherited the disease. On the other hand, the family could also have four children all of whom would be severely affected.

Contrasted to the recessive diseases are the dominant diseases. In these conditions the normal gene cannot overcome the effects of an abnormal gene in a gene pair, and the child inheriting the gene becomes the victim of the disease. If only one parent is a carrier of a dominant gene, the statistical treatment indicated earlier shows that the probabilities for combinations are for two pairs to contain one normal and one abnormal gene, and for two pairs to contain two normal genes. It would be predicted, therefore, that there would be a 50 percent chance that each child born of these parents would inherit the autosomal dominant disease.

Thus, the importance of knowing the genetic makeup of a couple planning to have children. A defective gene may be transmitted for generations without harmful effects as long as only one parent transmits the gene, and as long as the gene is recessive. However, as mentioned above, if the gene is present in both parents there is a statistically predictable chance that the resulting child will be the victim of the disease.

Nature has been cruel in dealing with some individual members of the human species. All people have not been created physically and mentally equal. And we ourselves have contributed much to making "natural man" unnatural.

The National Foundation—March of Dimes has compiled statistics that indicate the serious nature of the problem. About 7 percent of all babies born annually in the United States, about 250,000, have defects detectable at birth or within the first year of life. Of these, about 140,000, or 4 percent of the total, show significant defects such as sickle cell disease, cystic fibrosis, etc., as contrasted with less important defects such as birthmarks and extra fingers or toes. The Foundation further estimates that about 20 percent of all birth defects are caused solely by genetic factors, an additional 60 percent by a combination of genetic and environmental factors, and the remaining 20 percent by environmental factors alone. Thus, a total of 80 percent of the defects would have some genetic cause.

About 1.2 million infants, children, and adults are hospitalized annually as a result of birth defects. About 15 million living Americans have birth defects that affect their daily lives. The Foundation also estimates that such defects claim about 560,000 lives each year—60,000 children and adults, and 500,000 who die before birth.

There are at least 2,000 distinguishable human diseases that are genetically determined. Of these at least 92 have been shown to be caused by a genetically determined specific enzyme deficiency.

Most individual genetic diseases are rare, the average being about one in every 10,000 live births. However, the large number of such diseases brings the total of those afflicted to over 5 percent of all live births, or roughly one in twenty. The many genetic diseases may be divided into three general types: those due to chromosomal abnormalities, those resulting from the defect of a single gene, and those resulting from defects of multiple genes.

Any modification of either the structure of a single chromosome or the number of chromosomes is considered a chromosomal abnormality, although all such abnormalities do not result in physical or mental defects. Such aberrations occur about once in every 200 live births, but undesirable side effects result only if a functional gene is involved.

Chromosomes can be visualized with an ordinary light microscope, but the individual genes cannot. It is possible, by a procedure known as karyotyping, to arrange the 23 chromosome pairs into a pattern, based on size, shape, and, since 1970, on the property of each chromosome to absorb stains or dye in different, specific ways. Thus, when chromosomes are released from a cell, spread onto the surface of a slide and photographed, it is possible to cut out the reproductions of individual pairs of chromosomes and arrange them in a now standardized pattern. Chromosome pairs are labeled 1 through 22, with the sex chromosomes, XX or XY, being the 23rd pair. Such a final, organized photograph is called an ideogram. Thus, by observing the ideogram it is possible to detect abnormalities in chromosome shape or number, and to identify which chromosome pair is involved.

Perhaps the most familiar variation from the normal number of chromosomes is trisomy-21, three chromosomes at position 21. This one extra chromosome produces Down's Syndrome, formerly called mongolism, a condition that results in the commonest form of institutionalized mental retardation.

What causes these abnormalities is still one of the unsolved mysteries of science. There is some relationship between the incidence of Down's Syndrome and probably chromosomal abnormalities in general, and the age of the mother. For example, for a mother under the age of 30 the incidence of Down's Syndrome is about 1 in 1,500. Between the ages of 30 to 34 this rises to 1 in 1,000, for ages 35 to 39 it is 1 in 300, for 40 to 44, 1 in 100, and over 45 the incidence is about 1 in 40. The risk of any chromosomal variation shows the same sharp increase in occurrence with age, with the incidence being about twice the Down's Syndrome incidence for any age group.

Attempts have been made to explain this age-incidence relationship by associating it with the age of the fertilized ovum that develops into the afflicted child. Each woman is born with all the ova that she will release over the rest of her life. Thus, the egg of a 40-year-old woman is 20 years older than the egg released by the same woman when she is 20. Many geneticists believe that something in this aging process results in the observed abnormalities.

As might be expected, because of the great diversity of traits normally controlled by the genes there is also the same diversity of effects exhibited when a gene becomes defective. Since genes control every physical and biological aspect of bodily function, including appearance, it should be no surprise that, even though we can attribute a class of diseases generally to "single gene defects," resulting diseases are radically different. Sickle cell anemia and Huntington's Disease are examples of diseases caused by a single gene defect, but differing in practically every aspect of the resulting symptoms. Sickle cell anemia is prevalent among black populations, and rare among whites, and is caused by a defective gene that produces an abnormal hemoglobin known as hemoglobin-S. The disease derives its name from the fact that the red blood cells of the victim have a crescent or sickle shape, rather than the usually flat disc. Sickle cells die faster than normal cells, so a person acutely afflicted will develop severe anemia. Both men and women may carry the defective gene.

Sickle cell disease is an autosomal recessive disease, that is, it is not determined by a sex gene, and, as indicated by the term recessive, the abnormal member of the gene pair is neutralized by the normal member. It is only when an offspring inherits the abnormal gene from both parents that severe clinical symptoms occur. Individuals who inherit the gene from only one parent are usually symptom-free or, at worst, have only a mild case of anemia.

There is an interesting sidelight to this disease. Those who have inherited only one gene from a parent are more resistant to malaria. This creates an ideal situation for a natural propagation of a genetically deficient population. In malaria-infested countries, those who are resistant to malaria live longer and are more healthy than others. Natural selection arranges for those with mild sickle cell anemia to increase in numbers. It has been reported that as many as 30 percent of the natives of some malaria-infested regions of Africa have sickle cell anemia in some form.

Huntington's disease, or Huntington's chorea, is a disease characterized by involuntary jerking movements and by mental deterioration. Personality changes such as moodiness, obstinacy, and lack of initiative sometimes appear as the first symptoms before the lack of muscular control becomes evident. As the disease progresses, judgment and memory become worse, the patient may become irascible and destructive, walking becomes impossible, and mental deterioration becomes profound. It is a particularly tragic disease because the symptoms usually do not occur until about the age of 35 to 45. As the situation exists now, the victim lives a totally symptom-free life until middle age with no indication of the tragic end in store. This sense of security persists well into and sometimes past the time when children would be produced. The disease is autosomal dominant, so that each offspring has a 50 percent chance of inheriting the disease.

Recent work indicates that Huntington's Disease may be diagnosed as early as twenty years before the onset of symptoms by testing with laevo-DOPA. This could be both good news and bad news. It would give the potential victim the chance to make a decision concerning the advisability of having children, but it would also result in twenty years of apprehension while waiting for symptoms to appear.

The term "inborn errors of metabolism" is used to describe a genetic disorder that prevents an individual from metabolizing some particular chemical in the normal way. Without the ability to metabolize or break down the chemical, it builds up in the body, usually causing undesirable physiological reactions. About 125 known birth defects are attributed to this cause.

Phenylketonuria (PKU) is probably the best known of these diseases. It results from the inability of the system to form the enzyme that oxidizes phenylalanine, an amino acid, to tyrosine, another amino acid that is excreted normally. With high concentrations of phenylalanine present, some is converted to phenylpyruvic acid. The diagnosis of the disease involves the determination of abnormal concentrations of phenylpyruvic acid in the urine. Since phenylpyruvic acid is chemically a phenylketone, the disease is commonly known as phenylketonuria or PKU. About one in 10,000 to one in 15,000 children are born each year with the genetic defect. If undetected, it causes severe mental retardation. However, since both the detection and the cure are simple, the disease is not the same threat that it once was. Treatment consists of maintaining a diet that is low in phenylalanine.

Some common birth defects such as cleft palate, club foot, pyloric stenosis, some types of congenital heart disease, have some genetic etiology, but are probably caused by the malfunctioning of more than one pair of genes. Thus, it is impossible to make any statistical predictions as to the likelihood of a child of an affected parent having the same condition.

The list of conditions goes on—achondroplasia, a form of dwarfism occurring in one in 37,000 births; cystic fibrosis, with an incidence of one in 2,000 births; homocystinuria, one in 160,000 births; maple syrup urine disease, one in 170,000 births; cystinuria, one in 16,000 births.

It is interesting to note the number of diseases associated with specific ethnic populations, or that differ in incidence between populations. Tay-Sachs' disease, a condition that causes deafness, blindness, and deterioration of the nervous system and results in death usually by the second or third year, is limited almost entirely to those tracing their ancestry to the Ashkenazi Jews of central and eastern Europe. It is an autosomal recessive disease, but is also classed as an inborn error of metabolism, in this case lipid metabolism.

Sickle cell anemia is concentrated in black populations. Thalassemia, or Cooley's syndrome, a disease causing anemia because of the inability of the body to produce a normal amount of hemoglobin, occurs in those of Italian or Greek descent.

Club foot occurs once in 240 births in the white population of the United States while the incidence is one in 355 among blacks. Cleft lip and palate is present once in every 750 white births, but only once in 2,450 black births. An extra fifth finger occurs in one in 700 whites, and one in 80 blacks.

There is no reason to believe that there is anything inherently different between the races that would cause different genetic effects to occur at different rates. The differences can be rationalized on a demographic and environmental basis.

If because of some natural mutation, a genetic variation occurred in some geographical area, it is reasonable to assume that it would be passed on to future descendants of those living in the region. Unless there was some intermarriage with others in different locations the disease would be limited to natives or descendants of natives of the region where the original variation occurred. Thus, a disease that is limited to one ethnic group would be observed, not because the group happened to be Jewish or Greek or Italian but because the genetic variation originated in those particular ethnic societies.

It is apparent, too, that the environment plays a large part in determining whether or not a genetic variation will produce undesirable results. For example, children suffering from galactosemia, a disease that prevents the metabolizing of milk, will have no symptoms unless they drink milk. Similarly, a PKU sufferer will be symptom-free if no phenylalanine is ingested.

An interesting hypothesis has been proposed for the prevalence of hypertension in the black population of the United States where blacks are twice as likely as whites to develop this condition. Dr. Richard Gillum of the University of Minnesota has proposed that blacks may have inherited a kidney system that was well suited to the rigors of African life but, when exposed to the United States environment, is ineffective. In Africa, a highly effective mechanism for retaining body salt is very desirable because of a tendency to lose salt through perspiration in that climate, and because the normal diet is very low in salt. However, when exposed to the high salt level diet and the moderate climate of the United States, an abnormally large amount of salt would be retained, thus contributing to an increase in blood pressure.

Thus we see the methods by which genetic defects are transmitted, and the wide diversity of the diseases resulting from these defects. Without some intervention there is no hope for an actual reduction in the number of those affected by genetic diseases. It is variously estimated that every living individual carries between four and eight defective genes although, because of some of the special circumstances necessary for the propagation of a genetic trait already discussed, including environmental factors, the resulting incidence of defects is not the huge number that might be predicted on a statistical basis. However, as long as these aberrations are present and as long as reproduction takes place, genetic diseases will exist.

In fact, even if we assume that there is no change in the types of mutations or genetic defects that occur, the numbers of carriers of these defects are bound to increase. With the same proportion in the population, numbers have increased proportionately to the population. However, of even greater significance is the fact that carriers of defects and sufferers of a severe disease caused by the defect now, because of better medical care and treatment of symptoms, live longer and reproduce other carriers. As an example, before it became possible to detect and treat infants suffering from phenylketonuria, sufferers very seldom lived a life that allowed them to produce children. With the advent of effective treatment these people now live normal lives and, although showing no symptoms, are still carriers of the defect and pass it along to their offspring. In other words, before the days of modern medicine, natural selection limited the total numbers of individuals as carriers by preventing reproduction.

Hemophilia, a condition where the defective gene is carried by the mother but where the disease occurs only in males, is another example of a disease that can now be kept under control, albeit very expensively, with medication (anti-hemophilic globulin in the form of special cryoprecipitate). Motulsky calculates that, with normal mortality and fertility, the frequency rate of the disease will increase from one in 14,000 to one in 7,000 in 100 years. A woman who carries this recessive X-linked gene has a 50-50 chance that her son will have the disease.

In the case of pyloric stenosis, an operation at birth preserves the life of a newborn, thus also preserving the carrier of a genetic disease.

Totally aside from the emotional suffering caused by these conditions, the economic cost is staggering. Average estimates by various experts put the percentage of hospital beds occupied by persons suffering from conditions either wholly or partly genetic in origin at 25 percent. The cost to the United States of caring for only one group of affected persons, those suffering from Down's Syndrome (with a frequency of about one in six hundred births), is estimated by the National Foundation to be $1.7 billion annually.

According to H. V. Aposhiam, 36 million future life years were lost in 1967 from birth defects, primarily genetic, in the United States.[1] This is four and a half times as many future life years as those lost from heart disease, eight times as many as cancer, and ten times as many as stroke.

The problem, then, is huge. But the benefits resulting from a successful conquest of genetic diseases are also huge. Benefits to the individuals, even to nations, will result. The course of history could be changed in the future by making people more comfortable in their health environment. The challenge is apparent. Can we correct the inequities that nature, sometimes stimulated by our actions, has visited upon us? Is it now possible to think of a time when every child born into the world is guaranteed, not just the right to live, but the right to be born with a sound physical and mental constitution based on a sound genetic background?

In future chapters I shall discuss how this may be done in many cases with techniques that are already available, techniques that allow the kind of genetic changes that have taken nature eons to accomplish, to be made now in a generation, a year, or even a month.

But that is not the end of the challenge. Can we bring about these changes without degrading ourselves in the process? I have described the terrible consequences of a wide range of genetic diseases. I have indicated that there are even methods of eliminating or preventing some of them. The obvious conclusion might be that whatever technical knowledge is available should be used to solve this univeral problem that causes untold suffering both to the victims of the diseases and those closely associated with them, and that causes almost incalculable economic losses around the world. Even though methods do exist to solve some of these problems should the methods be used?

We should be aware not only of what can be done, but also the physical, ethical, and moral consequences of using the tremendous power available to us to intervene in natural processes.

4 . . . *wherein it becomes possible to change the blueprint of life*

Covering one wall of the Camera della Segnatura in the Vatican is Raphael's famous painting, *The School of Athens*. It was commissioned by Pope Julius II shortly after 1500 to commemorate the status of science. In the picture we see Socrates counting off on his fingers the propositions of a syllogism. We see Alcibiades who was a student of Socrates and whose only other claim to fame was that he was a traitor to three different countries, one of them twice. We see Pythagoras who taught the transmigration of souls and the sanctity of all life, and who was interested in mathematics and the theory of music, and Epicurus who taught that the supreme test of truth was to be found in the sensations, that there were only natural causes. We see Heraclitus, who considered fire as the first principle of all things, and Diogenes the cynic, and the geographer Ptolemy, and the mathematician Euclid and even the astrologer Zoroaster.

The two central figures are Plato and Aristotle, Plato with his finger extended upward to indicate his belief that universal ideas originated in heaven, and Aristotle, with his hand extended between heaven and earth to express the idea that universal ideas are formed by abstraction from the perception of the senses. I was so impressed with both the beauty of the painting and its significance that I have reproductions hanging in both my office and my home. Its significance for me lay in the fact that it is a dramatic example of the relationship of modern science to ancient philosophy. In the 16th century a painting commemorating *science* includes those people we recognize as *philosophers*. But on reflection it is obvious the difference between the "natural philosophers" and present scientists is that today's scientists perform experiments, while the natural philosophers tried to solve problems by reason alone.

It is probably not unusual then that, since the difference is in the approach to a problem and not the problem itself, many of the problems considered by the philosophers are still the problems of science. Aristotle, describing the ideal state in his *Politics,* says: "As to the exposure of children, let there be a law that no deformed child shall live. However, let no child be exposed because of excess population, but when couples have too many children, let abortion be procurred before sense and life have begun."[1] Aristotle's statement does not differ too greatly from that of Bentley Glass, made 2,400 years later: "No parent in that future time will have a right to burden society with a malformed or mentally incompetent child—they must produce a man who can transcend his present nature."[2]

The two statements express a common objective: to eliminate the birth of deformed or mentally handicapped children. While the objective is the same, the methods of arriving at it can be completely different, because of our ability to investigate rather than just hypothesize. In ancient times there was no thought of preventing or curing deformities, because there was no idea of what caused the deformities. They were curses called down upon parents, they were the revenge of the gods, they were the inevitable result of living and reproducing. Now, we can at last believe that methods other than exposing a newborn child to the elements or eliminating a fetus before birth by abortion can be developed to solve the problem.

Human beings, by nature, are optimistic. Hope is ever present in them. So with the publicizing of the advances of genetic manipulation, genetic engineering, and genetic screening, the expectations of the world have been raised to the point where it is believed that human physical characteristics, even intellectual abilities, can be changed and improved by the almost miraculous techniques being developed.

Since the recognition of the importance of DNA in genetics, advances in the field of genetic manipulation have been staggering. Characteristically, we have been almost euphoric in postulating the benefits that might come from these new facts. What are some of the results now being obtained in this miraculous area of science and medicine? They read like realized dreams of those occupants of research labs of only three decades ago—dreams that were thought even that short time ago to be almost science fantasy.

By a relatively simple process known as amniocentesis, it is now possible to give to the human fetus what amounts to a medical check-up after about the sixteenth to the eighteenth week of pregnancy. By means of a needle inserted into the womb, some of the amniotic fluid surrounding the fetus can be extracted. The fluid contains some of the cells shed from the fetus during early pregnancy. Analysis of this fluid makes it possible to detect some 70 metabolic disorders as well as all known chromosomal abnormalities. It is also possible to determine the age, weight, sex, and state of lung development of the fetus. Using amniocentesis, many genetic defects present in the fetus could be determined. If the defect would result in a serious mental or physical problem for the newborn, abortion, if acceptable, could be elected. If there is no detectable defect, the parents could be reassured that their child would be born free of conditions for which the test had been performed. If newborns are screened, remedial procedures could be initiated immediately.

Withdrawal of amniotic fluid from a pregnant woman for therapeutic purposes has been known for more than a century. The first really useful diagnostic results came as a result of the work of Dr. C. A. Bevis in England and Dr. A. W. Liley in New Zealand on the Rh factor in 1952. In 1956 F. Fuchs reported on his work identifying sex chromatin in the nuclei of cells in the amniotic fluid. Eleven years later, Dr. Carlo Valenti of Brooklyn and Dr. Henry Nadler of Chicago independently used this technique to diagnose Down's Syndrome in the fetus.

Late in 1975 the Department of Health, Education and Welfare approved the use of amniocentesis for pre-natal diagnosis. The decision was based on a four year study of more than 4,000 women sponsored by the National Institute of Child Health and Human Development. In the 1,040 women who had amniocentesis performed, matched by 992 controls, the accuracy of the diagnosis was 99.3 percent. Among the women showing positive results, 19 had fetuses with genetically caused metabolic disorders. There were six false results in the 1,040 women tested. Two babies were born with Down's Syndrome after the test had shown they would be normal. In one case a fetus was diagnosed as having galactosemia, but was born normal. There were three cases where the sex was incorrectly identified.

The National Foundation—March of Dimes reported that, in 1974, 2,187 women in 37 institutions had diagnostic amniocentesis during their second trimester of pregnancy. Of these, 2,125, or 97.2

percent, were carrying babies diagnosed as normal. At a 1978 meeting of the American Association for the Advancement of Science it was estimated that amniocentesis had been performed on 20,000 women in 1977.

Presently, the only thing medicine can offer a mother carrying a defective fetus is abortion. Thus, the whole technique of amniocentesis is objected to by anti-abortionists, on the grounds that it encourages abortion. On the other hand, the National Foundation points out that definite knowledge of which fetuses are defective and which are not might actually decrease the number of abortions performed. In cases where the parents carry genetic defects, they might be inclined to elect abortion rather than take a chance of having a defective child, even though the odds are in favor of the child being normal. With the assurance of normalcy, the parents could be released from doubts caused by statistics alone, and thus would allow the child to be born.

Amniocentesis performed for the purpose of deciding whether or not an abortion should be performed does have one disadvantage. The procedure should be performed between the fourteenth and the twentieth week after the last menstrual period. It then takes some weeks for the laboratory procedure to be run, so that, by the time a decision to abort can be made, the fetus is probably sixteen to twenty-one weeks old at the very earliest.

There are several indications for the selection of women to undergo amniocentesis. One would be if either parent is carrying a defective gene; another would be maternal age. The risk of bearing a Down's Syndrome child increases significantly as the mother grows older. It is estimated that there are 300,000 pregnant women over thirty-five years of age in the United states each year, and they produce half the children affected with Down's Syndrome. Reproductively active people within ethnic groups susceptible to specific genetic disorders such as Tay-Sachs' disease or parents who have had a previous child with a genetic disorder would also be candidates. If all these standards really do indicate candidates for amniocentesis, this would amount to as many as 400,000 women per year.

Obviously, the solution sought by scientists to the elimination of genetic defects is not abortion. It must be either prevention or cure. For example, it has been suggested that cystic fibrosis could be completely eliminated over the next forty years by screening all pregnancies and aborting the 17 million fetuses that will carry a single gene for this disease. Dr. Leon R. Kass of the Committee on Life Sciences and Social Policy, National Research Council, National Academy of Sciences has said: "Such zealots need to be reminded of the consequences should each geneticist be allowed an equal assault on his favorite genetic disorder, given that each human being is a carrier for some four to eight such recessive lethal genetic diseases."[3] However, the first step must be identification and, as long as the procedure is safe and accurate, amniocentesis is justified as a first step.

I have here discussed amniocentesis as it relates to a technique used for genetic evaluation. However, it has also been an extremely valuable procedure for assessing the degree of fetal involvement in Rh incompatability, and also in determining fetal lung maturity, gestational age, and fetal distress.

Our discussion so far has been limited to the possibility of identifying and predicting genetic abnormalities, or of gathering pre-natal information. However, there does exist a possibility of altering genes at some stage during the development of an animal body.

It is fascinating to consider the specificity of genes. Each individual has in his or her genetic makeup thousands of genes. Yet specific gene pairs or combinations of gene pairs are responsible for one child being born with Down's Syndrome, another for a child having sickle cell disease, another for a child having phenylketonuria, and so on for the hundreds of conditions known to be caused by gene defects. But the exciting part is that each specific disease is caused by similarly specific, identifiable, detectable, genes—always the same, always carried, and always transmitted. It is at this point that the imagination can visualize results that would be of invaluable benefit to humanity. First of all, if these defects can be detected in a carrier, then the results of reproduction could be predicted. With prediction could come the means of prevention.

If defects can be associated with specific genes, perhaps they could be cured. In other words, after a child is born, or even in the womb, the makeup of that particular gene might be changed to transform the individual into a normal organism.

Before discussing the possibility of eliminating genetic diseases or influencing human characteristics by the control of genes, a word of caution is in order. Only some diseases are caused by single, identifiable genes. In other words, the diseases are monogenic traits. However, by far the greatest number of traits are not determined by a single genetic factor but by constellations of genes. Additionally, genetic traits (mono or polygenic) are known also to be influenced by environmental factors that can be a major consequence in the appearance of symptoms. For example, a person carrying the gene for Xeroderma pigmentosum is perfectly normal until exposed to sunlight, when generalized skin tumors occur. Those who carry a defective X-linked gene for an enzyme involved in metabolizing sugar (G6PD) may show no symptoms of a disease until they are exposed to some drugs and foods, including aspirin and the broad bean *Vicia faba*—then they contract hemolytic anemia.

For those monogenic identified defects there can be, at least conceptually, some hope for cure. The basic assumption would be that, if a gene is defective, it should be replaced. The incorporation of an externally supplied gene into a human cell, or the controlled modification of a specific gene, is a formidable task. Two processes are theoretically possible. The first is called *transformation,* and involves the direct transfer of DNA. The second is *transduction,* and involves incorporation of blocks of genes by means of a non-lethal virus invading the cell, carrying with it the desired gene and incorporating it into the cell.

The difficulties seem almost insurmountable. But considering the incredible progress made in developing techniques for the manipulation of genes, of chromosomes, of DNA and RNA, for the identification and determination of the structures of tremendously complex molecules, and even the synthesis of genes, hope returns.

Among the early workers in this field was Dr. Heinz Fraenkel-Conrat, who worked with the tobacco mosaic virus, so called because the virus produces a mosaic pattern on tobacco leaves after infecting the cells. A virus is composed of an inner core of nucleic acids, covered along its length by a protein coat. Fraenkel-Conrat removed the protein coat from the virus and showed that the coat itself could not infect the tobacco. He then recombined a new core of nucleic acid with the old protein coat and obtained an infectious virus. Finally, he combined the core nucleic acid with the protein coat obtained from a mutant of the virus, and again obtained an infectious material, indicating that it was the core that was the determining portion of the molecule.

It was mentioned earlier that the absence of the enzyme transferase causes galactosemia. If, in some way, a genetic change could be made to take place such that the enzyme would be produced, the disease could be eliminated. Dr. Carl Merrill and his associates at the NIH in 1971 succeeded in growing skin cells that had milk-metabolizing properties. Using the technique of tissue culture, which means growing cells such as skin, kidney, etc., in an artificial medium in the laboratory, he was able to infect skin cells with a virus, lambda phage. The virus was incorporated into the cells, and the genetic information from the enzyme-producing virus was transmitted to the cells and became part of the genetic characteristic of later generations of cells.

It is a tremendous step from skin cells in a test tube tissue culture to the elimination of a genetic disease in a human being. However, the principle has been proven. Dr. Allen S. Fox, Professor of Genetics at the University of Wisconsin, succeeded in altering the genetic characteristics in an organism above the level of the bacteria, the fruit fly. Ordinarily the egg of the fruit fly is covered by a chemically-resistant membrane. Dr. Fox has been able to remove the membrane without damaging the egg. He then immersed the unprotected eggs in a solution of DNA prepared from other flies possessing the gene that he wanted to transmit. By this means he was able to modify the eye coloring of the resulting fruit flies.

Further up the evolutionary ladder, Dr. Beatrice Mintz has combined the cells of two mouse species into a single species. She took the blastocysts, that is, the earliest stage of division of a fertilized egg, of two species at the eight cell stage, dissociated and rearranged them into an eight cell blastocyst composed of the mixed cells of the two species. The blastocyst was allowed to develop *in vitro* and was then implanted into a female mouse, where it developed normally. Thus, the resulting mouse could be said to have had four parents.

The Chimera—half man, half animal—has been famous in myths throughout the ages. Now, the concept of developing a Chimera is no longer limited to myth. Henry Harris, professor of pathology at Oxford, has succeeded not only in combining the body cells of mice with those of humans, but in keeping them alive and reproducing. Since the common cell has the mixed chromosomes of both mice and human, it also has the same hereditary characteristics. If the same kind of combination could be achieved with the reproductive cells, the ovum or the sperm, a living body, truly a Chimera, could be developed.

Whether or not genes can be transferred from a higher organism, a eukaryote, into a lower organism, a prokaryote, and still function has been a question that has remained unproved, in spite of the speculation over the years concerning just that result of genetic manipulation. In 1976 two laboratories independently proved that this is possible. At Stanford University a segment of DNA isolated from bakers' yeast, a eukaryote, was inserted into the prokaryote, *E. coli*. The resulting *E. coli* functioned as a normal bacterium possessing the properties transferred to it from the yeast. At the University of California, Santa Barbara, similar experiments were performed and, in addition, preliminary evidence obtained showed that genes from the fruit fly can also function in *E. coli* bacteria.

Another advance in the gathering of basic information on which to found further research on genetic manipulation came when Nobel laureate Har Gobind Khorana at MIT announced the synthesis of a totally artificial gene that could be inserted into a cell with resulting normal activity of the cell. The synthetic gene corrected a mutational defect by replacing a defective gene in a virus, bacteriophage lambda.

Earlier, in 1970, Khorana, while at the University of Wisconsin, had synthesized a gene for the first time. While it was a great scientific triumph, this gene could not be studied further since there were no methods available for determining whether or not it could be transferred to and function in a cell. The new gene, announced in 1977, when incorporated into the proper organisms, functioned to allow the system to produce protein.

The significance of this work is not so much that it allows the correction of a genetic defect in a bacteria or a virus, but rather that it opens up methods for the study of the mechanism of gene action. If gene structures can be changed at will, the effects can then be studied in test tubes or in cell systems. Such results in lower organisms should be considered tremendously encouraging. But the application of these techniques to the prevention or cure of a genetic defect is a different matter. Consider the requirements for this application.

First, it would be necessary to work only with a specific gene, or with groups of genes, that would give only the specific desired results. Second, it would be necessary to have the gene incorporated into all the cells requiring it, and to eliminate an undesirable gene, if it were present. Finally, it would be necessary for the cells to reproduce with the incorporated gene as a component of the cell. In order to have genes introduced by transduction, the system to introduce the genes would have to penetrate the cell wall and then the wall of the nucleus within the cell. The new gene would have to be protected from degradation by the cell, and it would then have to be carried to the correct position on the chromosomes. The viruses seem to be logical candidates for carriers, since they do penetrate cells and nuclei, and since they are able to pick up DNA.

Another method proposed for repairing or eliminating defective genes or of introducing genes with different genetic properties is by means of chemical mutagens. It is known that many chemicals can cause mutations in organisms. However, the problem involved in using chemicals in any controlled manner to modify a specific gene in humans is truly stupendous. The difficulties arise from at least two different circumstances, both relating to problems of experimentation and prediction.

First, if preliminary work is to be carried out in animals, a chemical may be effective in one species and not in another. Further, it may be active in males and not in females, or vice versa, and, to make matters worse, it might even be effective in one hybrid of the species but not in others.

The second difficulty involved in these chemical mutagens arises from the near impossibility of measuring results in human beings. To determine whether a chemical is causing mutations in humans would involve a carefully controlled test of huge numbers, estimated by statisticians to be at least twenty million people. Related to this is the time factor. Most genetic mutations involve recessive genes. The effect on an offspring becomes apparent usually only if both parents have the same recessive gene, and if they both pass it on to the same offspring. Therefore, effects might not show up for generations—possibly hundreds of years, when, by chance, both parents possessed this gene.

In a single animal test to determine the mutagenicity of a single chemical, triethylenemelamine, conducted at Oak Ridge National Laboratories in 1965–66, about 24,000 offspring of 500 male and 500 female mice were studied over a period of eighteen months. It has been estimated that to obtain the equivalent data in human beings would require the study of half the world's population for a million years.

Most chemical mutagens attack only one of the four different nucleotide bases of DNA. There is no way of knowing where the attacked base will be in any gene and, consequently, there is also no way of knowing what specific chromosome will be changed. An effect on a gene or a chromosome will affect one of literally tens of thousands of activities involving enzymes or proteins. If the action cannot be predicted, it is almost impossible to determine the effect of the mutagen on an individual or in a population if the only method of doing this is by observation of individuals in the population. Since many thousands of different effects might be produced, no single effect will occur often enough to be judged significant, particularly in view of the fact that the same mutation might occur spontaneously as the result of environmental factors such as sunlight.

The same difficulty is encountered in determining whether or not chemicals present in the environment are causing genetic effects. One of the methods proposed as a gross method for determining general bad effects is to monitor the male–female birth ratio. In the United States, 105 males are born for every 100 females. This ratio exists in spite of the fact that the male, in the prenatal and early stages of life, is indeed the weaker sex. The higher number of males to females born is explained by the fact that more males are conceived. It is estimated that as many as 120 males are conceived for every 100 females, thus allowing more male embryos to die while still maintaining the 105 to 100 ratio.

Two advances in technology have allowed research in the field of genetic manipulation to take a giant step forward. The first of these is the ability to "sequence" DNA, that is the ability to determine the exact order in which various bases occur along a chain in a DNA molecule. If the exact structures of parts of the molecule are known it might be possible to determine which are the active sites on the molecule and thus the requirements for specific activity. Only a few years ago it took teams of research workers two years to determine the sequence of a part of a DNA molecule containing twenty bases. Today the same sequence can be determined in one day.

The second breakthrough was the discovery of "restriction enzymes," that is, enzymes that could cut genes sharply and at very specific and known positions. The enzymes, produced by bacteria, recognize a specific DNA sequence and cut the DNA at that sequence. Each different enzyme recognizes a different sequence, so that various parts of a DNA molecule can be separated with certainty. At least eighty such enzymes are now known. Much of the early work on restriction enzymes was done by Dr. Herbert W. Boyer at the

University of California. When a gene is cut, the tendency is for the cut end to recombine with some other active fragment present. Thus, if two cut ends are present in the same medium, they could combine.

It is obvious what possibilities this opens up for modifying genes. If the modified gene is incorporated into a cell, all daughter cells will then carry the traits of the modified gene, whatever they may be. If a modified gene should be incorporated into a bacteria such as *E. coli,* that divides every twenty minutes, billions of new genes would be produced every twenty-four hours. The potential, both good and bad, from such developments is staggering. On the good side, one of the most obvious hopes is that of attacking diseases caused by genetic defects. With the ability to manipulate genes, defective genes could be eliminated, or new genes capable of producing desired traits could be substituted.

If the human traits that allow us to manufacture biologically important substances could be transferred to bacteria, for example, it would be possible to make these substances in the laboratory. If higher organism DNA that manufactures insulin in the body could be incorporated into a bacteria, the bacteria would then be a factory for the production of insulin. The same concept could be applied to the production of hormones and antibodies, thus eliminating the necessity to innoculate an individual with a live or inactivated virus or bacteria for vaccination.

This area of research is already producing encouraging results at a speed much faster than even enthusiasts had predicted. In 1977 a group headed by Dr. William J. Rutter and Dr. Howard M. Goodman at the University of California, San Francisco, succeeded in transferring insulin genes from a rat into the bacteria *E. coli*. Thus, the potential for converting growing bacteria into an insulin-manufacturing plant.

The most dramatic achievement so far has been the insertion of an artificial gene fragment ''code'' into a special strain of *E. coli*. The bacteria so produced, when grown in culture medium, produced somatostatin, a human hormone that controls the body's production of insulin and growth hormone. The work was reported by Dr. Boyer at the University of California. The growth of 100 grams of bacteria in 2 gallons of culture medium produced about 5 milligrams of the product. The hormone, normally produced in the hypothalamus at the base of the brain, had been isolated previously for the first time in milligram quantities after the laborious extraction of ground brain tissue of half a million sheep. Thus, there is now unequivocal proof that genetic information can be transferred from a higher to a lower cell, and still act in a useful manner.

Only months after Boyer's achievement, scientists at Harvard University and the Joslin Diabetes Foundation of Boston announced that they had succeeded in making a rat insulin gene function in *E. coli*. The resulting organism produced proinsulin, a protein easily converted to insulin. Although rat insulin cannot be used by human beings, the technique used is also applicable to the production of insulin from cattle or pigs, the two sources of insulin for diabetics at present, or eventually even human insulin.

Harvard has applied for a patent on the process for producing insulin, and has indicated it will grant licenses in return for the funding of further research. In late 1978, scientists at City of Hope National Medical Center, Duarte, California, and Genentech, Incorporated, a private research company in San Francisco, announced the production of insulin that is identical to that produced in the human body.

Insulin is now extracted from the pancreas of animals, specifically swine and cattle. It requires about 16,000 pounds of pancreas glands to produce 2.2 pounds of insulin that in turn will take care of about 1,650 diabetics for one year. About 1.5 million people are presently receiving insulin injections to control their diabetes.

The potential uses for genetic engineering range beyond the medical field. Already many laboratories around the world are seeking to modify organisms to carry out various industrial operations. So far, the organisms used for these purposes have been those that have occurred as mutants in nature. For example, organisms are known that will preferentially bind certain metals. By proper engineering, an organism could be developed to separate almost any metal, including platinum, gold, and silver, from sea water, sewage plants, or even from the basic ores. Considerable work has been done on the conversion of petroleum to protein.

General Electric has been successfully working on an organism that will digest oil. A bacteria has been developed that will metabolize many of the components of oil. It has been suggested that this could be a solution to cleaning up oil deposited in oil slicks when they occur.

The control of cross breeding in agriculture, both with animals and plants, could be speeded up immeasurably if genes could be transferred at will into the plant or animal under investigation, rather than waiting for the generations now necessary to be produced by nature. In cattle, grass is metabolized and converted to energy, not because of any genetic trait peculiar to cattle, but because their stomachs contain bacteria that convert cellulosic foodstuffs to utilizable food. No such bacteria exist in humans. However, with genetic manipulation techniques it might be possible to transfer to *E. coli* those characteristics necessary for cellulose metabolism. Since *E. coli* readily adapts to the human intestinal system, it is possible to postulate that people could then eat grass as a nourishing food.

Plant life has also been proposed as a logical target for genetic manipulation. Root nodules of some legumes have the ability to fix nitrogen from the soil. If this trait could be transferred to other crops, wheat and oats for example, the need for fertilization would be reduced drastically.

To go to the extreme, some animal traits might be transferred to man. The DNA in the genetic codes of plants and animals is similar in structure to human DNA, except that it is in different sequence. Thus, genes that allow animals to hibernate might be useful for astronauts on long space missions.

But for every benefit proposed for gene modification, an equally great risk has been postulated. When modified genes are incorporated into an organism, a new organism results. Opponents of gene research point out that it is impossible to predict the properties of such an organism, or the effects it might have on a human population. For example, an organism capable of causing a new or even old disease, resistant to antibiotics, might be created. The organism might cause cancer, with results observable only after fifteen or twenty years. Work on the organism that metabolizes oil has been discontinued because of the fear that, once released on oil slicks, there would be nothing to prevent its taking over entire oil wells and destroying oil wherever it occurs. Lethal genes, uncontrollable by present methods, could be introduced into plants and animals.

For those who fear the mutations possible from genetic manipulation it should be pointed out that mutations are the stuff from which evolution is made. Were it not for mutations in nature many of the millions of species that inhabit the earth would not be in existence. If there is something to be feared it is the speed at which evolution can now possibly take place, mere *months* instead of millions of years. We are now at a place where we can change evolution, including our own future. It would be reassuring to think that our wisdom has increased with our ability.

It is the compression of time in which these changes can take place that causes doubts and fears. True, changes have taken place in nature, but they have taken place in small increments of change, each tested by time, most eliminated if they are deleterious. Now it is possible to go from the means to the total end, usually irreversibly, without any testing of the consequences or, in some cases, a knowledge of what the consequences will be. The factor that really causes apprehension is that whatever effect is obtained is irreversible. If an undesirable organism is created, it could reproduce forevermore. If an undesirable gene trait is expressed in individuals, the trait will be passed from the parents to the children and to their children's children.

A second sobering fact is the simplicity of the technique. Almost anybody with a laboratory can now carry out recombinant DNA experiments, with the possibility of creating new life forms. The combination of the potential dangers of such research and the knowledge that many laboratories, either so unskilled or poorly equipped that proper safeguards could not be substituted, could carry out this work prompted some researchers to take an action unprecedented in scientific history. In 1974 a group of the most outstanding scientists working in the field of genetic research called on their fellow workers to evaluate their own work, and to refrain from gene modification in those areas where the resulting product might be more dangerous than the original organism.

As a result of this public call to action, a worldwide conference attended by 140 scientists, as well as nonscientists, met at Asilomar in Pacific Grove, California, in February 1975 to consider the question. Both opponents and proponents of the continuation of such research presented their positions, and the subject was vigorously debated. The consensus was that such research should be continued, although many individuals still disagree. Dr. Jon Beckwith, the Harvard scientist who performed pioneering work on bacterial genes, has refused to participate in any gene transplantation experiments.

The feeling that gene transplantation could unleash deadly new microbes forced the cancellation of this research at the Institut Pasteur in Paris. More than 300 researchers, technicians, and teachers at top French institutions joined to protest gene-splicing research.

The World Health Organization's Advisory Committee on Medical Research, however, recommended that such genetic research continue. They believe that the risk can be sufficiently minimized to justify the continuation of gene research. But the Committee did recommend technical studies on safer equipment and techniques.

The decision to proceed with new methods of genetic manipulation was not reached without considerable trauma, soul-searching, and admirable submission of personal ambitions to scientific progress on the part of the participants in the decision. Following the Conference of Asilomar, a National Institutes of Health Committee met to prepare a set of rules to guide experiments in recombinant DNA. At a meeting of the Committee in July 1975 at Woods Hole, the recommendations were discussed and seriously weakened. The intense criticism this weakened version received resulted in a second subcommittee being appointed, and a new set of recommendations being drawn up. These new rules would be so strict that some of the experiments actually in progress in the laboratories of the participants of the decision would have to be eliminated. In general, the guidelines evaluate each experiment, and define the precautions that must be taken during its progress. Precautions are defined in two different areas: the environment in which the work is taking place, and the living organism that is used. There are four levels of laboratory safety precautions labeled P1 to P4, with P1 being described as standard microbiologic practice, and P4 requiring stringent controls. Among these are: the experiment must be performed in a laboratory under positive pressure with air locks; workers must wear protective clothing and take showers before leaving the laboratory, and others. Similarly, there are three danger levels for the organism used—EK1 to EK3, with EK1 requiring the experimenter to use the standard K12 strain of *E. coli,* while EK3 requires that the organism used be proven by experiment unable to survive at a level of one bacterium in 100 million outside the laboratory. Experiments are classified by various combinations of the above limitations, for example P2 + EK3, or P4 + EK2.

After two years of debate, the NIH published its final guidelines in June of 1976. The guidelines do not differ substantially from those agreed to earlier by the NIH Recombinant DNA Committee.

In spite of the careful consideration that went into developing the standards under which recombinant DNA research can be conducted, and in spite of the fact that they have been agreed to by the top scientific arm of the U.S. government, doubt still remains as to how successful they will or can be in controlling such research. One huge loophole exists. At present there is no way to enforce them. The guidelines apply now only to those doing research under NIH sponsorship, and the control consists in withholding or granting funds. In March of 1977 the National Institute of Health was funding 123 research grants involving recombinant DNA, the National Science Foundation was sponsoring fifty-two, and the Veteran's Administration was responsible for eight. At least eight pharmaceutical companies announced programs in this field.

Other bureaucracies soon began to seek their own controls. At Cambridge, the City Council held open hearings to consider whether or not Harvard should be allowed to do gene-splicing research within the city limits. On June 23 and again on July 7, 1976, the Council held hearings on the subject. The discussion could only be described as intense. The faculty of Harvard split, with members testifying on each side with equal vigor. The mayor of Cambridge wanted to know, "What's going to crawl out of the tubes?" If the decision were left with him he would "ban the research in Cambridge, period." The Council, after the second meeting, declared a three-month moratorium on research that required a high or a moderate level containment laboratory (levels P3 or P4 in the language of the regulations), and established a citizens' committee to examine the issue. The committee itself was composed of nine Cambridge residents, some of whom had little or no science background, a fact that caused the research workers some uneasiness. On February 7, 1977, the Council accepted a report from its Citizens' Committee. It recommended that research be allowed to continue under the NIH guidelines, but with some additional restrictions. Research at the P4 level, that requiring the highest level of physical containment, was banned. If P3 work is undertaken it must be done with a disabled (EK2) organism to prevent the growth of the new organism outside the experimental environment. Finally, all laboratories doing P2 or P3 research must be free of rodent and insect infestation.

This is the first instance of a non-scientifically knowledgeable committee involving itself in a highly technical policy matter. It will not be the last.

The difficulty in establishing rules for laboratory experiments in genetic manipulation lies in the fact that there is no real information available on the extent of the hazards involved. In imagination all the horror stories could come true. Totally new and deadly organisms could be produced and escape from the lab, infecting all beings on the earth with uncontrollable disease. In fact, new organisms never before known to have occurred in nature will be produced. I say "known to have occurred," since it has been postulated that possibly thousands of these organisms have come into being, but have remained undetected because they were not able to survive. One of the arguments used by the proponents of gene research is that just such occurrences have taken place, and that if they have not survived in nature, they should be safe in the laboratory.

It has been assumed that nature does not permit a creation of a new organism resulting from the combination of a higher with a lower organism. In scientific terms, there is a barrier between eukaryotes, the higher organisms, and prokaryotes, the lower. However, it is stated that since there is every opportunity for this to have happened, it probably has happened and is happening. For example, it has been estimated that our population excretes from the human gut 10^{22} or ten billion trillion organisms every day. In the event that human DNA split in some way, and came into contact with a bacteria, for example *E. coli* in the gut, an organism composed of human–bacteria traits would result. The chance of this happening could be statistically very small, but, because of the enormous number of bacteria always available, the statistical chance of its happening would be significant. The fact that such organisms have not been detected is attributed to the fact that natural selection eliminates them.

As might be expected, Federal Regulations were proposed immediately as a solution to the problem. It is unrealistic to believe that every worker who can do this research—those in academic institutions, industry, and research institutes, as well as independent investigators—will abide voluntarily by the NIH guidelines. Bills were quickly introduced in both the House and the Senate to require a license to do gene-splitting research. After a year of debate and consideration the original bills were sidetracked in favor of a continuing Congressional study of the situation. The change in attitude came about primarily as a result of rational presentations of scientists, some of whom had originally been strongly in favor of eliminating research in this field. It is still probable that some legal control will be instituted. It is hoped that the same wisdom, both from scientists and nonscientists, will be applied to future control as has been applied to the preliminary guidelines.

The present policies are aimed at eliminating perceived dangers resulting from recombinant DNA experiments. The proposals seek to do this by controlling one technique, namely gene-splitting and recombination. The same hazards might result from other experiments such as cell fusion, mutagenesis, and recombination methods other than gene-splicing. Thus, eventually, the control would have to be any technique resulting in a new DNA, not just on a specific technique.

In formulating policies to limit research in certain fields it must be recognized that the problem is international in nature. Laboratories all over the world, at varying levels of competence, are capable of carrying out a whole range of genetic experiments. It would do no good to prohibit dangerous experiments here in the United States if even one area of the world permitted them. It is unreasonable to suppose that all countries would agree to our strict guidelines. For example, at a meeting of the NIH committee on recombinant DNA held in December 1977, Dr. John Tooze of the European Molecular Biology Organization, West Germany, said, "Our view is that the U.S. guidelines are naively overstringent." Several other European observers echoed these sentiments. In most cases ability to perform experiments depends strictly on the competence of the experimenter, not the location of the laboratory. It is quite conceivable that, in the absence of worldwide prohibitions, scientists working in prohibited fields who did not agree with the prohibition would gravitate to the centers where the work was still permissible.

The objection to this new and seemingly exotic research is another indication of an increasing tendency to seek a risk-free existence for society. It can't be done. Certainly, we can never prove that it *can* be done. By definition, it is impossible to prove a negative. In a lifetime or a series of lifetimes it could not be assumed that anything—a drug, an automobile, a food—would not cause a bad effect in some way. One can only gain more and more experience, and try to extrapolate that experience to safety considerations. Every time a physician administers a drug to a person an experiment is being performed. In spite of the millions of people who might have received the identical drug with complete safety there can be no assurance that this particular person might not react differently.

We already have some experience with recombinant DNA. It is not all that new. For more than six years—at least since late 1973—DNA from viruses, frogs, insects, and various other species have been incorporated into *E. coli* without incident, in some cases with less containment than that now prescribed under the NIH guidelines. Probably billions of bacteria have been exposed to this technique.

It is easier to dream up possible calamities than it is to devise experiments to show that the dream can never be a reality. The dream of the risk-free world is just that—a dream.

René Dubos once said,

> A society that does not continue to grow through adventure and willingness to take chances is not likely to survive long in the modern world. With regard to health as to all other fields, society must be willing to take educated and calculated risks because they are inherent in technological civilization.[4]

Today, as in the past, willingness to take risks is a condition of biological success.

Almost five years after the meeting at Asilomar the tone of the debate has become much more reasonable. To be sure, shrill voices are raised defending extreme positions—from "eliminate all research in this area" to "absolutely no restrictions are necessary." Some of the most eminent researchers in the field, some of those who first stimulated the moratorium, after further study of available data and in view of constantly developing new information, have changed their positions as to the possible dangers. Even the Congress and Senate have withdrawn from their former position that legislation was necessary, and that only new laws could protect the public.

In new guidelines published in the July 28, 1978 *Federal Register,* several significant changes over those proposed originally were included. Fears had been expressed that the *E. coli* strain used might become pathogenic and adapt itself to humans. The new regulations express confidence in the safety of the procedure, and eliminate the requirement that experiments be performed only in the two highest grades of safety, P4 physical containment and EK3 biological containment. Under the new guidelines gene-splicing experiments with animal virus DNA may now be conducted under P1 and P2 levels of containment, rather than P3 or P4 as originally required.

The requirements for performing shotgun experiments, experiments where the entire gene set of an organism is broken into segments, have been reduced from the highest level of containment to P2, which is not much more than a standard laboratory.

One of the interesting sidelights that has developed during the several years of discussion has been the open break between activist environmental groups and scientists formerly associated with the objectives of these organizations. Paul Ehrlich, one of the most prominent and most active environmentalists of the scientific community and a trustee of Friends of the Earth, wrote to the director of the NIH saying it would be foolhardy to restrict research based on imagined risks. Louis Thomas, president of the Sloan-Kettering Cancer Center, resigned from the same Friends of the Earth board in protest over their activities. The Natural Resources Defense Council (NRDC) finds itself in a similar difficulty. Joshua Lederburg, a former Nobel Prize winner, president of Rockefeller University, and a trustee of NRDC, dissociated himself from the position taken by that group. René Dubos, another Nobel laureate, also protested NRDC activities in an angry letter to the director of NIH. It read in part:

> You may assume that I had been consulted about the preparation of this document and that I am in favor of its recommendations. But this is not the case. I had no idea that NRDC was involved in the recombinant DNA problem, for which it has no competence. . . . The failure on the part of NRDC to communicate with me . . . reveals either irresponsible lack of familiarity with the literature in this field, or intellectual dishonesty in using my name for a cause that I regard as ridiculous.[5]

Other scientists, both those involved in DNA research and those outside this research area, have expressed the feeling that those advocating the prevention or strict control of DNA research are wrong, and that their efforts, now without the support of prominent scientists, are doing a disservice to the environmental movement.

Finally, in a Solomon-like effort to satisfy all sides, the HEW, in late 1978, further revised its "final" guidelines. This time the announcement of the revision was made not by the NIH, the body responsible for supervising the research guidelines, but by the Secretary of HEW. Along with the revised standards the Secretary asked the Environmental Protection Agency to insure compliance with the guidelines. The revised guidelines also require that 23 percent of the members of local institutional bio-safety committees at each local institution that review and approve research before it is started, be representatives of the general public.

The new guidelines do exempt from regulation five categories of experiments that comprise about one-third of recombinant DNA research covered under the original guidelines, and that the NIH has concluded do not present known health risks.

The extent of the progress in this area, and the degree to which the technology is being accepted, is indicated by the fact that the U.S. Court of Customs and Patent Appeals ruled that a micro-organism that had been genetically modified, in this case the organism that digests oil, could be patented. A previous decision in which approval was given for patenting a live organism for the production of an antibiotic was used as a precedent. The court stated there was no reason to refuse patent protection "to micro-organisms themselves—a kind of tool used by chemical manufacturers in much the same way as they use chemical elements, compounds, and compositions which aren't considered to be alive." Although the oil-devouring organism was not produced by recombinant DNA techniques, the resulting organism is in the same category as those that will result from gene-splitting. It is, indeed, a startling change of attitude when a living organism is described simply as a "tool" and is compared to non-living chemicals.

My own conclusion is that there is practically no chance of a calamity befalling the human race as a result of any research that has been proposed so far. On the other hand, there are immediate benefits apparent. I do not believe we will see for years to come the free exchange of genes in the human body, the curing of genetic diseases. I do see the gaining of a tremendous amount of basic information that will sometime make this possible.

In the meantime, let's get on with it.

5 . . . wherein a conflict develops between the ability to detect genetic defects and the desirability of such detection

Starting with what is now possible we can say that techniques are available for detecting many genetic defects in any human being whether it be an adult, a newborn child, or even a fetus. Knowledge of this capability should result in rejoicing and dancing in the streets. The determination of the genetic characteristics of potential parents makes it possible to decide on the risk of conceiving a child who might inherit an undesirable characteristic present in one or both parents. *In utero* or *ex utero* determination of genetic defects in infants make possible the early diagnosis and, in a few cases, treatment of severe diseases.

But nothing is simple. The potential for evil seems to be equal to the potential for good, at least in the minds of a large and vocal part of society. Some see genetic screening as a way to collect information that could be used as a method to cause genocide in certain populations. If all blacks who were sickle cell carriers were prevented from having children, the result would be a tremendous drop in the black population. Others see it as a way of further stigmatizing certain individuals with the classification of "handicapped." If genetic information were used as a basis for employment, discrimination against anyone not considered "normal" could be practiced.

The proponents and opponents of mass genetic screening can be divided roughly into three groups: those who believe mass screening for all identifiable and testable defects should be mandatory, those who believe no screening should be done, and those who believe in mass screening but on a totally voluntary basis. The doubts raised about such mass programs revolve around the uses of the information obtained. Will it be used in beneficial ways or will it be used to control or even eliminate populations? Or will the data be useless medically and bothersome emotionally?

There are no curable genetic diseases in the sense that the genetic defect can be eliminated. There are very few diseases in which those afflicted could be improved by prior knowledge. PKU and galactosemia would be two of these, since proper treatment could be instituted as soon as the afflicted child is born. In the absence of the possibility of cures, then, two other positive actions could result from the use of information. A determination could be made *in utero* of those fetuses with defective genes and a decision made as to whether or not to abort the fetus, and parents with defective genes could be counseled as to the consequences of having children.

The great fear motivating those who oppose genetic screening is that the information obtained from the screens would become available to other than the individual and his or her physician with dire consequences to the individual. For example, might a prospective employer discriminate against someone who is a sickle cell carrier? Could an insurance company insure an individual who was a predicted victim of a debilitating disease, even though the prediction could not be verified for many years into the future? Such questions are being asked by those who describe genetic screening as an invasion of privacy.

Inevitably, at the best, public pressure would be brought to bear not to produce children who might become a burden to society, just as pressure is being applied to the poor now. At the worst, restrictive laws might be passed limiting the benefits of those who did insist on the privilege of reproduction.

How should mass screening be set up and justified? Sickle cell anemia is a disease limited almost entirely to black populations. Should the entire population be screened for this trait? PKU is rarely found in blacks. Should blacks be excluded from mass screening? Tay-Sachs' disease is limited to Jews. Should non-Jews be screened?

It would seem logical and reasonable in these cases to eliminate from screening those populations in which the disease is rare. However, the emotional issues involved cannot be overlooked. Already there is the general opinion that minority ethnic groups and the poor are discriminated against in the matter of medical care. Any program that was confined only to white populations would encounter considerable opposition. On the other hand, programs instituted primarily for blacks have been opposed by the black populations themselves because they see the results of these screening tests as threatening genocide. Sickle cell disease is a case in point. It is well known that this is a disease limited to blacks, that there is a high incidence of the trait carried by the black population, and that there is no cure for the disease. Therefore, the only way to reduce the incidence is for those individuals who are carriers to refrain from having children.

It is difficult not to attach some social stigma to those carrying defective genes. The Boston XYY case is illustrative of the problems involved. There is one pair of sex chromosomes among the twenty-three pairs in each individual. In the woman, the chromosomes are alike, both X. In the man there is an X chromosome and a Y chromosome. In a very small number of men there is an abnormal chromosome pattern that consists of an extra Y chromosome. In other words, they have an XYY chromosome pattern instead of XY.

In the early 1960s, studies of inmates of mental-penal institutions showed that about 2 percent had the XYY chromosome pattern. In newborn infants studied at the same time the incidence was only 0.11 percent. On the basis of this difference the tremendous jump was taken to the conclusion that the presence of the XYY pattern was associated with aggressive or sexual psychopathology. Based on the incidence of XYY in the general population and on the prison population, it can be calculated that as many as 97 percent of the XYYs will never end up in mental-penal institutions. From this it must be inferred that the vast majority will lead normal lives, even if we assume that some who should be institutionalized are not.

However, because of the interest in the problem and the publicity attending it, research groups decided it would be important to determine experimentally whether or not the XYY chromosome did indeed influence behavior. In 1968 Dr. Stanley Walzer, a Harvard psychiatrist, and Dr. Park Gerald, a geneticist at the Boston Hospital for Women, established a program aimed at screening all male infants to detect three different kinds of chromosome patterns: those with an extra X chromosome (the XXY or "Klinefelter's Syndrome"), those with the extra Y, the XYY pattern, and those with the normal XY arrangement. Plans had been proposed to follow the development of the children for at least twenty years, with detailed psychological and behavioral studies being conducted on each child during that period.

However, after the program had been in progress for some time, vigorous opposition developed. What had started as a routine testing of each newborn male infant evolved into a major program. Originally, consent forms were not required from the parents. However, eventually consent forms were required, a full description of the program was provided to the parents, and they were told that they would be informed if chromosome abnormalities were detected.

In June of 1975 Dr. Walzer announced the discontinuance of the screening program, giving as his reason the emotionally exhausting atmosphere created by the controversy. The objections to the program were vigorous, and are typical of the objections to many research programs on humans. In the first place, it was stated that the determination of the presence of the XYY pattern could not be considered predictive, although that was the question Dr. Walzer was trying to answer twenty years hence. More seriously, however, the objectors questioned the results on the parents of the child, with the resulting influence on the development of the child. Suppose the XYY pattern were found. What effect would this have on the reaction of the parents to the child's behavior? If they believed that the child had a tendency to aggressiveness, would they over-react to normal aggressive responses, and thus adversely affect the activity? As the child grew older and became aware of his condition would he assume that aggressiveness was predestined, and thus become aggressive? Additionally, would the XYY condition be considered a stigma? Would employers discriminate against someone they thought might, in the future, exhibit undesirable behavior?

Finally, the program was criticized on its scientific merits, with the objectors saying that no objective results could be obtained since both the experimentor and the parents knew which children were XYY, and this, in itself, would influence results. However, if there really is an association between aggressive or sexual psychopathology and the XYY, then parents could be counselled in how to deal with these episodes. As long as there is doubt about the association, there will be questions about the influence of counsel on the outcome of the experiment. The dilemma is that, unless a newly designed experiment is performed, the association can never be proven. I suppose it could also be considered a benefit if the study ended with negative results, that is, that a lack of association were proven. At least the minds of parents of XYY children would be at ease.

The fear that minority groups, as well as others, have of the misuse of genetic screening has some basis in fact. Consider the Illinois Domestic Relations Act drawn up by the Chicago Bar Association. It provides that all applicants for marriage licenses be tested for venereal disease and for "any other disease or abnormalities causing birth defects." Additionally, the Association suggested that, in the future, the state should "require that a stated abnormality peculiar to certain races, for example sickle cell anemia (affecting primarily black people) or Tay-Sachs' disease (affecting primarily Jewish people) should be corrected as a condition preceding the issuance of a license." It is no great compliment to the Association to point out that they were at least reasonable enough to state that, at the present time, "it is impossible to impose such a requirement."

In 1972 Maryland passed a law that directed that information about sickle cell disease and the availability of tests and counseling be provided to all couples seeking marriage licenses. There was nothing compulsory about the bill. It simply assured that information be supplied, and that information clearly distinguish between the trait and the disease. The bill ran into considerable criticism on the basis of the fact that it discriminated against one group of people, namely blacks. The state then came up with a new bill that provides for the screening of a wide variety of conditions.

The District of Columbia and the Assembly of Virginia adopted mandatory screening laws for children prior to elementary school admittance. To do this legally, it was necessary to declare sickle cell anemia a communicable disease, an obviously ridiculous conclusion. Virginia also requires Hemoglobin-S (HbS) testing for marriage applicants and inmates of correctional and mental institutions. One wonders why such screening is mandatory, particularly in correctional and mental institutions. What purpose is served? Is it a prelude to, for example, mandatory sterilization? Requiring this test for engaged couples has a degree of cruelty attached to it, since the decision to marry has already been made.

The problem is not even as simple as conducting a screen, informing individuals privately of their condition, educating them as to the possible consequences of having children, and then allowing them to make an informed decision. A surprising number of people simply do not want to know of an incurable or non-preventable condition.

We have already seen several abortive attempts to develop a

screen that will detect cancer at an early stage. The question has been asked seriously whether or not it would be of benefit to try to develop such a test. Suppose, as has happened, a trial is made of a test that will detect cancer of any kind before there are any symptoms of the disease. Suppose one million people were screened in the original test. Since one out of four people now living will contract cancer at some time in their lives, it would be expected that 250,000 people would give positive results in this test, assuming that a cancer predisposition was always present in the individual. For the sake of conservatism, however, let's assume that only 125,000 are far enough advanced to be detectable, but these are detected in symptom-free people. From the time of the first positive result, the individual will probably live an anxiety-filled life waiting for the first symptoms of cancer to appear. The one advantage of having this knowledge is that, in some cases, the earlier the cancer is detected, the more hope there is for successful treatment. However, in many others, early knowledge would make no difference.

Another factor to be considered in any screening test is the chance of error. In any biological screen, a test that gives no more than 5 percent false positives and 5 percent false negatives would be considered an excellent test. In the case we have chosen, then, over 6,000 people would spend their lives in apprehension because the test had predicted they would be cancer victims, when all the time they should have been free of worry. On the other side, the same number of people would be reassured, and would ultimately develop symptoms. Expand this screening to entire populations, and the enormity of the problem becomes apparent.

Most practitioners take the attitude that the test should be oversensitive—that it is better to have a false positive than a false negative. There is a certain logic in this position, particularly if the screen is used as just that, a screen, to identify those individuals who should have a further follow-up confirmatory test to diagnose the disease. The position breaks down when treatment of the disease is based on the screen, and as much harm can come from the wrong treatment as can come from the disease itself, as in the case of PKU. If a test having a 99.99 percent accuracy is used to detect a disease occurring once in 20,000, then there would be one false positive found for every real positive detected. This means that there would be the same error in using the screen as there would have been if no screening had taken place. In these cases, as in all others, cost must be balanced against the benefits. The costs are the years of apprehension and fear in which many people will live. The benefits are the savings of life and suffering of those who are benefited by early detection.

Is it better not to know?

There is also the question of what effect the knowledge of the possibility or probability of future disease might have on an individual, particularly when there is nothing that could be done about the condition. For example, there is evidence to support the belief that it is possible to predict whether or not an individual would develop Huntington's Disease, that usually strikes its victims during the middle stage of life. Would twenty years or so of anticipation be more than the prospective victim could bear? And is this justification sufficient to prohibit this test from being run? Many ethicists say the test should not be run—it is better not to know. Others point to the fact that for every individual exposed to this trauma there would be thousands who would be reassured because the test had shown them to be risk-free.

Finally, there is the responsibility of the potential victims to want to know. Should they, for their own sake as well as others, not have the duty to decide whether or not to assume family responsibilities when evidence points to the fact that they will sometime suffer from a destructive and burdensome incapacitation? In knowing, individuals cannot spare themselves. However, they can spare suffering to any others.

The question then arises as to what information is to be given to those requesting genetic tests. As an example, a pregnant woman, worried about the possibility of Down's Syndrome in her child, requests amniocentesis. The test was negative for Down's Syndrome, but instead of a normal XY male chromosome pattern, the abnormal XYY pattern was detected. Should the mother be told about this abnormality, since she had not requested such a test and was relieved by the negative findings for Down's Syndrome, particularly in view of the doubtful significance of the results?

An even more sensitive area is exemplified by a case in which a couple were being advised about their child who had developed sickle cell anemia. The wife volunteered the fact that it was her fault, since only she and not her husband had the sickle cell trait. This was confirmed. The couple obviously did not understand the genetics involved. The counselor wisely did not explain, since it would immediately have indicated that the husband was not the father of the child.

At least forty-three states have programs for the screening of newborn infants for PKU. It is estimated that 90 percent of all the babies born in the United States are screened before they leave the hospital. As more information becomes available more states are expanding their PKU programs to include other genetic conditions.

PKU is probably the ideal test to use to develop programs for genetic screening. The test, developed by Dr. Robert Guthrie of the State University of New York in 1961, is simple and inexpensive. Several drops of blood are taken from the newborn, dried on a filter paper disk, and mailed to the testing lab. There the disks are placed on a plate containing agar, bacteria, and a specific bacterial-inhibiting substance. The bacteria will grow only if a predetermined minimal amount of phenylalanine is present (two mg per 100 ml). If an increased zone of bacterial growth is observed on the test plate, the infant is suspected to have PKU and follow-up tests are conducted. The initial test has been automated to the point where one technician can test about 50,000 samples per year.

PKU occurs often enough to see tangible benefits from detecting the afflicted infants, and there is a treatment available for the disease. At present about three million babies are screened each year. About three thousand of these will give positive results, of which about two hundred will have classic PKU. Treatment consists of putting the child on a low phenylalanine diet for as long as six years. "Diet" is actually too gentle a term to use for the restrictions on what the child can eat. Practically no food can be eaten with the exception of a preparation called Lofenalac, which is a powder that is mixed with water before feeding. The problem this causes a family might be imagined. The preparation costs about $60 per month. But the treatment is successful, and is the alternative to developing mental retardation.

The apparent success of the PKU screen has developed a false sense of confidence in those now proposing expansion into other diseases. For the most part, new screens have been added to programs without much regard for the logic, reasonableness, or benefit of the screen, and without any general overall planning. For example, the District of Columbia screened 77,000 babies at a cost of $135,000 over three years without finding a single case of PKU, which is quite rare in blacks. The mandatory program has now been repealed.

Rhode Island screens for maple syrup urine disease, a disease with an incidence of one in 170,000 births. It would be expected that a case would be detected about once in every five years in this state.

The cost of these programs is not high, at least by other medical standards. New York has screened every infant born in the last two years for seven different conditions at a total cost of $250,000 per year. Massachusetts tests between 75,000 and 80,000 infants per year, and detects some metabolic disorders in 30 to 35, of whom about 60 percent develop a clinically significant disease. The cost is about $2.50 per infant screened.

Maryland is the exception to my statement that most of the programs are developing with little or no overall planning. In 1973 a Commission on Hereditary Disorders was established with the responsibility for deciding what screens are to be included in their program and how they are to be administered. Significantly, the charge to the Commission included the provision that the public should have some influence on the type of program developed. It was specified that there should be provisions for consulting services, for the preservation of confidentiality, and for the protection of the rights of those screened. Specifically, one principle forbade any restriction on childbearing. Originally, the Maryland PKU screening program was mandatory. The law has since been repealed, and all screening is now voluntary.

It is easy to demonstrate the economic advantage of having a screening program to detect curable diseases such as PKU. If 10,000 were screened at a rate of $5.00 per infant (this is a high figure) and if one PKU were found, the entire $50,000 could be said to have been expended in finding that one infant. However, suppose there would have been no screening program. The child would almost certainly have ended up institutionalized for an average of fifty years. If the cost were only $5,000 per year (a low figure) the total cost to take care of the individual would amount to $250,000. Even this figure does not include the cost that could be estimated due to the loss of productivity of the person.

After two and a half years of study the National Academy of Sciences released a report on genetic screening. Several significant attitudes resulted from these deliberations. They reported favorably on genetic screening, but under carefully controlled circumstances. They emphatically opposed any laws or regulations that make screening mandatory. Finally, and of great significance, they believe that communities should be involved in any decision regarding screening, and that there should be citizen participation to make the program effective.

The Institute of Society, Ethics, and the Life Sciences has published a committee report on genetic screening. They proposed three goals for screening programs:

1. Provision of benefits to individuals and families;
2. Acquisition of knowledge about genetic disease;
3. Reduction of the frequency of apparently deleterious genes.

As principles to be applied to the design and operation of screening programs they specified the following:

1. Attainable purpose
2. Community participation
3. Equal access
4. Adequate testing procedures
5. Absence of compulsion
6. Informed consent
7. Protection of subjects
8. Access to information
9. Provision of counseling
10. Understandable relation to therapy
11. Protection of the right of privacy

In general, I believe the advantages of mass genetic screening far outweigh the disadvantages. An awareness of the potential dangers of misuse will help to minimize them.

The problem involved with the interpreting of genetic screening data, and the explanation of its significance to involved individuals has given rise to a new profession, the genetic counselor. The data showing the results of genetic counseling are still very scarce. In a study conducted in England involving 421 couples, forty percent of the couples facing a one-in-three chance of having a defective baby had at least one pregnancy after the counseling. In the same group the divorce rate was three times the national average. Counseling, in these cases, changed the expectation of the involved couples for having babies, resulting in a feeling of personal failure. In a case reported from another clinic, one family decided to have a child even though there was a 100 percent chance that offspring would have a fatal genetic disorder.

I question the need for a special category of counselors. The only thing special about the situation is that those requiring counsel have genetic problems. The emphasis, therefore, should be on being a counselor, rather than on being a geneticist. The basic knowledge of genetics required to counsel in this area is easily obtained. The skills required to be a successful counselor in any area are many, and those possessing them are rare. It is easier to learn the facts of science than to learn the combination of patience, compassion, firmness, objectivity, sympathy, appreciation of individual rights, and the recognition of individual dignity required when trying to give a person all the information necessary to allow him or her to make a decision affecting life and death.

There are also legal implications of counseling advice. It is not uncommon for those suffering the results of some genetic accident to consider a third party responsible for their condition. The following four examples are illustrative of what can happen.

During her pregnancy the mother of Jerry Gleitman contracted German measles, with the result that Jerry was born with serious defects in sight, hearing, and speech. Jerry brought suit in New Jersey against the physician involved on the grounds that the doctor had not advised his mother to have an abortion, and thus prevent his birth as a defective person. In New York a child conceived as a result of rape in a New York mental institution sued the state on the theory that it should have prevented his conception by mentally deficient unmarried parents. In Illinois a bastard child sued his father, arguing that the father wronged him by allowing him to be born. In 1977 in Springfield, Illinois, the Illinois Supreme Court ruled that a child could sue her doctor and hospital for damages because of birth defects caused by a procedure performed on the mother nine years before the child was born. When the mother of the child was only thirteen she had been given a transfusion of Rh positive blood that was incompatible with her own Rh negative blood. She was not told of the mistake. The baby born nine years later required a complete change of blood, and suffered permanent brain damage.

Obviously, the attitude of the counselor can have a great effect on the decision made by the affected parties. It is easy to talk about objectivity, but hard to be objective when dealing with the great emotional trauma involved in most decision making relating to genetic abnormalities. Is it better to have lived abnormally for five or ten or twenty years than never to have lived at all? Is a Down's Syndrome child, capable of a loving life yet incapable of contributing to society, to be considered useless? Are the parent's feelings being given more weight than the rights of the child?

Marc Lappé has pointed out what he calls a western bias that many scientists bring to bear on ethical problems in science, and says he detects an overriding dependence on the ongoing ideal of the "greatest good for the greatest number." This has led some to overemphasize the future impact of having or not having children. He says genetic counseling may be misguided if it is felt that the counselor's ethical obligation is in any way to future generations.[2]

Acting out of compassion does not always mean that detection, for example, of a Down's Syndrome fetus automatically leads to abortion. A child with Down's Syndrome responds to human warmth and love. In whatever way happiness can be defined, these children can live happy lives. Certainly they are not unhappy. Parents of these children can and do respond to them with love, in many cases in a more responsive manner than to a normal child. In such situations, no suffering is added to the world, and the child certainly deserves to be born.

From the beginnings of medicine physicians have been faced with the problem of how much information to give their patients. Even today there is a sharp division among those practioners who say there are some details about an individual's condition that should be withheld for the patient's good, and those who say that the patient should be completely informed. The case of persons suffering from a terminal, untreatable disease causes the most anguish. Should the patients be informed of their hopeless positions, or should they be allowed to approach death without the mental suffering that would, in most cases, result from a forewarning?

The dilemma involved in whether or not to inform a person of a genetic defect is similar, in some respects, to that involved in other branches of medicine. In many instances an incurable disease is involved. In many instances, too, the prediction can be made some years in advance of death. There is one major difference for a large group of people with genetic defects: they will never suffer from the symptoms of a disease. The danger is not to themselves but to their offspring. There is no doubt about the fact that people have either the right to know or the right not to know that they or their children are carriers of a genetic trait, or that any one of them is of a genetic makeup such that a disease would become evident in the future.

Certainly there are special situations where it would be of no conceivable benefit, and might even be harmful, for the individual to have certain kinds of knowledge. I've already referred to such instances—the case of the woman who thought it only her fault that her child had sickle cell disease, and the case of the discovery of an XYY child when the significance of a trait might be misunderstood.

In general, then, I believe that genetic screening, combined with genetic counseling, is in both the public and individual interest. I believe that it must be voluntary, that it must be confidential. While an individual has a right not to know, I think he or she also has the responsibility to know. The knowledge of genetic variations cannot, in itself, be harmful. It can only cause damage when it is either ignored or misused.

6 . . . *wherein genetic variations occur in every human being, and some genes or individuals should be eliminated, changed, or tolerated*

It is fortunate that all variant genes do not protrude from human bodies as warts or boils or other imperfections. Conditioned as we are by the advertising media to what is desirable for physical attractiveness, a whole new standard would be required, since everyone in the world carries some defective genes. All genes are invisible. Very few deviant genes ever result in adverse effects. But in a culture that uses wigs to cover hair loss or to give the impression of different hair color, that uses contact lenses to change the color of eyes, that uses girdles to pull in, pads to push out, platform shoes to make taller, even the idea of a deviation from perfection caused by an invisible, nonfunctioning defective gene is troubling. It is this devotion to the concept of physical perfection that makes me uneasy when I hear discussions concerned with the elimination of variant genes. Genes that result in serious physical or mental handicaps are in one category. However, the mere presence of a defective gene, even though it will never result in an observable effect of any kind, is increasingly being looked upon as undesirable in our pursuit of physical perfection.

It seems an inevitable result of the advance of technology that, as our ability to screen for more genetic defects becomes greater and the screens are more widely used, our standards for considering what is normal will change. Before the days of genetic screening, an individual could carry any number of defective genes and, because there was no way of detecting these defects, there was also no reason to attribute any factor of inferiority to the individual. Now, with the ability to screen for hundreds of gene conditions, there is the chance that milder and milder conditions will be considered abnormal. If the view of society is that such people are abnormal, and are therefore in some way inferior, the effect will be tragic.

Every time I mention the word "defect" I do so with the knowledge that it is not a precise description of the situation I am trying to define. "Defect" carries the connotation of something wrong. Possibly a better expression would be genetic difference or genetic variation, since all genetic variations do not result in something harmful and, in fact, sometimes, as in the case of sickle cell in those exposed to malaria, might even be helpful. What we are really trying to express is a variation from all the genetic symmetry and uniformity predicted by the laws of nature.

If we define a genetic defect as a variant gene or chromosome that, if passed on to progeny, will cause an undesirable physical or mental disease, then we can begin to define who is genetically healthy and who is not.

But even this seemingly straightforward definition is subject to qualification. A recessive gene might be present in a man or a woman, and be passed on to generations forever without causing any undesirable disease. However, if the gene is present in both the father and the mother, then the undesirable effects occur. Since the number of these undesirable genes is astonishingly large and present in an astonishing proportion of the population, the problem must be viewed as tremendously complex. For example, Cavalli-Sforza and Bodmer in *The Genetics of Human Populations* state that almost every individual carries the equivalent of more than one lethal recessive gene. Evidence for this as a conservative estimate is contained in the study reported in 1956 on groups known to have a high percentage of consanguineous marriages. If two people are blood relatives, the chances of their being genetically similar, that is, having the same genetic makeup including gene defects, are greatly increased. Therefore, a man and a woman related by blood are more likely to have the same recessive genes, and so produce offspring exhibiting the disease caused by the gene defect. The closer the relationship, the more likely this situation will exist. For example, the marriage of first cousins will result in the appearance of the results of one-eighth of the total recessive genes carried by both parents. Obviously, it is impossible to conduct a controlled study to determine this effect in human beings. However, there are areas of the world, including the United States, where, because of geographical isolation or tradition, groups have been intermarrying. By studying the offspring of these marriages, it has been estimated that the number of defective genes resulting in a clinically serious disease is between three and five *for each individual.* Other studies have indicated a number as high as eight for defective genes in each individual. Is it any wonder, then, that there is difficulty in defining a healthy person, or determining what an effective screening program should be?

Knowing that such genetic variations exist probably produces a natural tendency to want to eliminate them. Before the knowledge of defective genes, even before the knowledge that genes were responsible for inherited diseases, the desire to "improve the race" by improved "breeding" was strong. One of the principal proponents of such racial upgrading was Sir Francis Galton whose theories were widely advocated in the late 19th and early 20th centuries. It was Galton who popularized both the word and the idea of "eugenics," that is derived from the Greek word meaning "well born." Galton believed that the qualities of the human race could be improved if "good" families were encouraged to have large numbers of children while "inferior" families were prevented from breeding. He states the aim of eugenics as being "to give to the more suitable races or strains of blood a better chance of prevailing speedily over the less suitable than they otherwise would have had."[1]

Even before Galton, Charles Darwin said:

> Thus the weak members of civilized society propagate their kind. No one who has attended to the breeding of domestic animals will doubt that this must be highly injurious to the race of man. It is surprising how soon want of care, or care wrongly directed, leads to the degeneration of a domesticated race; but, excepting in the case of man himself, hardly anyone is so ignorant as to allow his worst animals to breed.[2]

The increasing support for the eugenics movement was fueled in the latter part of the nineteenth century by an escalating class struggle between labor and management. Labor's dissatisfaction with wages and working conditions was emphasized by such reactions as the Haymarket Riot of 1886 in Chicago. An ideology that preached, as did the eugenics movement, that those who were unfit should not be allowed to breed, and those who were fit should, and that defined the upper class as fit or successful, and the laboring class as unfit or unsuccessful, was a ready-made mechanism for management to use to blame labor for all the ills of the country.

A further contributing factor to this feeling was the return to prominence of the Mendelian Laws of Heredity shortly after 1900. Since some human traits could be explained on the basis of the inheritance of single gene pairs, it was easy for many people to jump to the conclusion that all traits could be so explained.

As so often happens in movements such as these, the disciples of eugenics became much more radical than the originators. Dr. Harry Laughlin, of the Eugenics Record Office at Cold Springs Harbor, New York, became one of the most extreme of the eugenicists. He was a prime leader of the movement in the first quarter of the twentieth century. He favored not just the sterilization of individuals deemed to be unfit, but he extended his judgment to groups. Among those he defined as "socially inadequate" were criminals, epileptics, blind, deaf, or deformed persons, drunkards, drug addicts, tramps, paupers, orphans, and those suffering from chronic infectious diseases.

Between 1900 and 1924, the movement commanded considerable public support. In 1918 Margaret Sanger concluded that,

> All our problems are the result of over-breeding among the working class—more children from the fit and less from the unfit—that is the chief issue of birth control.[3]

Labor became even more of a target as the result of immigrants coming into the country in large numbers and joining the labor forces. In 1924 the U.S. government passed the Immigration Restriction Act to limit immigration from east European and Mediterranean countries. It was not one of the more glorious chapters in our history. Although both labor and business were strong proponents of the Act, support from and testimony by eugenicists was very influential in obtaining congressional approval. With no evidence whatever to support the view, Dr. Harry H. Laughlin, the principal advisor on eugenics to the House Committee on Immigration and Naturalization testified, "The recent immigrants, as a whole, present a higher percentage of inborn socially inadequate qualities than do the older stock."[4] The Immigration Restriction Act was not repealed until 1962.

It is unfortunate, but true, that an effort was made to give some scientific credence to this discrimination. In 1912 the U.S. Public Health Service commissioned a study to determine the incidence of feeblemindedness among immigrants. Using an IQ test as a base, the following incidences were reported: Hungarians, 83 percent, Russians, 87 percent, Jews, 83 percent, and Italians, 79 percent.

Between 1907 and 1931 thirty-three states passed compulsory sterilization laws applying to drunkenness, sexual perversion, and feeblemindedness, among other things. At least twenty thousand people were sterilized for these so-called crimes. In 1925 the Supreme Court upheld the constitutionality of the sterilization law in Virginia when it was challenged by Carrie Buck, an eighteen-year-old imbecile confined to a state hospital. The fact that Carrie's illegitimate child as well as her mother were also alleged to be feebleminded gave rise to Mr. Justice Holmes's oft-quoted remark in his majority decision: "Society can prevent those who are manifestly unfit from continuing their kind. Three generations of imbeciles are enough." Legal experts are fond of quoting this remark with the added information: "Mr. Justice Butler dissented."

Between 1915 and 1930 thirty-four states passed miscegenation laws prohibiting marriage between people of different racial groups. Some of these laws remained in effect until 1967, when the Supreme Court declared them unconstitutional.

By the late 1930s the ardor of the eugenicists in the United States had dimmed considerably. I think that the reason was not any lessening of a belief in the value and necessity of using eugenics to control the quality of a population, but was rather a recognition of the excesses possible in using a genetic approach such as was being preached in Germany at that time. During this period the word "eugenics" became closely associated with genocide and ambitions to develop, not just a high quality race, but a true "master race."

In the last quarter of a century there has been a renewal of interest in using eugenics to solve some of the perceived genetic difficulties. The major problem facing the human race, according to the geneticists, is the possibility of having so many defective genes in the gene pool that soon the whole race will become weak. In 1961, Theodosius Dobzhansky, the scientist-philosopher said:

> We are, then, faced with a dilemma—if we enable the weak and the deformed to live and to propagate their kind we face the prospect of a genetic twilight; but if we let them die or suffer when we can save them, we face the certainty of a moral twilight. How to escape the dilemma?[5]

Others are not as compassionate as Dobzhansky in approaching the problem. Joseph Fletcher, a theologian, said:

> We cannot accept the "invisible hand" of blind natural chance or random nature in genetics anymore than we could old Professor Jevon's theory of feast and famine in the 19th century laissez-faire economics based on sunspots and tidal movements. To be men we must be in control. That is the first and last ethical word. For when there is no choice there is no possibility of ethical action. Whatever we are compelled to do is amoral.[6]

Bentley Glass raises the question of obligatory control:

> When I read in the Bill of Rights of the United Nations that one incorrigible right of the individual is to reproduce, and that the right of every person to have a family is a basic human right that must not be infringed upon, I wonder whether this "right" is indeed to remain unrestricted. Is it not equally a right of every person to be born physically and mentally sound, capable of developing fully into a mature individual? Has society, which must support at great cost the burden of genetic misfortune resulting from mutation, chromosomal accidents, and prenatal harm inflicted by trauma or virus, no right at all to protect itself from the increasing misfortune? Should not the abortion of a seriously defective fetus be obligatory? Should not the loss of a defective child be recompensed by the opportunity to have another, a sound child by prenatal or postnatal adoption? Can we not devise laws and practices that will improve even though slowly the quality of our population while we retain individual choice and freedom to a great extent? Can not the institution of a greater freedom of choice in new respects compensate for the restriction of some time honored privileges?[7]

While recognizing the excesses of those who believe in the overriding importance of genetic influence on behavior, it should be recognized that there must be *some* genetic factor operant. There is the danger today of a backlash against conceding any importance to heredity as it influences behavior. This could be as grave a position as to say that only one's inherited genes determine how the individual will react. There are those who object even to research in this field because, they say, the results might be used for political reasons, reasons that might result in discrimination against certain segments of our society. However, most responsible investigators recognize the interdependence of both heredity and environmental factors in determining intelligence. Such recognition requires that characteristics of behavior be interpreted not just in the light of inherited genes, but also in the light of influence of the environment in which the characteristics are operating. With such a recognition the chance for discrimination is greatly reduced. To refer to the example given earlier in this chapter, it should no longer be possible to blame the activities of the early immigrants on their genetic makeup, but rather to consider the intolerable conditions under which they were forced to live and work.

The same fear of having the race "downgraded" is still with us today, but with a completely different set of data to consider. We now know how to predict in many cases the birth of those individuals feared as race contaminators in the past. The traumatic discussions today revolve around how to prevent such occurrences. The great difference now centers mostly on a changing attitude on individual versus societal rights. When a nation passes a compulsory sterilization law it clearly states it is protecting the rights of society at the expense of the individual. Now the question being asked is whether or not the state has such a right, regardless of the physical or mental conditions of the individual.

Clearly the state has exercised and still exercises the right to protect society in many instances. If a person is suffering from a dangerous and contagious disease, his or her activities are immediately restricted, with full approval of everyone. States still demand, as a prophylactic measure, that children be vaccinated against certain diseases before they are permitted to attend school. These are restrictions on human rights that the state should have. Now, however, we are talking about restrictions on two of the most precious rights an individual can possess—the right, first of all, to be born, and secondly, the right to reproduce. At a slightly lower level is the right of the individual to live as he or she pleases, to have the same opportunities for the pursuits of a livelihood, education, and happiness—all of which could be sacrificed by unwise interpretation and use of genetic information.

The early proponents of eugenics, including Galton, talked about races and strains. The objective was to "improve the breed." More recently mental and social characteristics have attained much more importance as goals for improvement. This change has come about partially as the result of the appreciation of the role that the environment plays in molding character and in determining the reactions of individuals. Hitler's leadership abilities might well have made him one of the great leaders of all time in another environment. His genetic characteristics would have been the same. Yet Hitler is naturally missing from every list of individuals whose genetic character should be duplicated or preserved. The decision on whether or not Hitler was good or bad was not based on his genetic characteristics. It was based on how he acted in his environment.

With the information at hand we can postulate the elimination or prevention of genetic defects by either positive or negative eugenics. Selective breeding for favorable genes is called positive eugenics. Therefore, Galton's encouragement of "good" families to have large numbers of children would be classed as positive. Now, however, Galton would have a different basis for determining what is "good." By means of genetic screening he could determine who had good genes and who had undesirable or "bad" genes. Whether this would satisfy him is doubtful, since one of his criteria, possibly his primary criterion, was whether or not an individual belonged to a desirable class. As we now know, nature does not limit good gene distribution only to those in the upper class, although it might be considered that many members of the lower class are there because of accidents of nature.

It is unrealistic to believe that positive eugenics can ever do much to improve the qualities of a race. The first difficulty is encountered in isolating any single quality or trait to be inherited. It is virtually impossible to affect any one inheritied characteristic without also affecting others. Therefore, there is always an opportunity for downgrading the complete organism, even though the single desired characteristic might be successfully propagated.

Negative eugenics seeks to upgrade the race by preventing whoever is defined as an inferior individual from being born. In the past eugenicists looked to sterilization as the ultimate preventive. Today, a knowledge of the genetic makeup of the parents makes it possible to counsel on the advisability of having offspring. If a defect is detected in the fetus, it may be eliminated by abortion. Or, as suggested in Chapter 4, the ability to manipulate genes might, some time in the future, make it possible to correct defective genes either in the womb or after birth. In Galton's time the emphasis was on improving the race. Eugenicists still fear, as has been indicated, the effect on the race, but the emphasis has now shifted to improving individuals as members of the race. I do not believe that either positive or negative eugenics can succeed in improving the quality of an entire society without some obligatory control. In those cases where government controls have been introduced, public reaction has been so great that freedom of the individual had to be restored. Today, the importance of genetic control is being appreciated more than at any time in the past because we have more knowledge about the mechanism than was available before. At the same time the importance of the rights of the individual is also being appreciated more. The purpose of genetic screening and of genetic counseling is almost entirely to allow for individual choice.

As long as there is individual choice, people with physical and mental deficiencies will be born. Our society has said that these people have a right to be born. Not only do they have the right to be born, they have the right to be accepted as full members of society.

If the problem of dealing with individuals with abnormal physical or mental characteristics cannot be dealt with by means of eugenics, it must, then, be solved by *euthenics*. Euthenics accepts the fact of abnormal or subnormal persons, but requires whatever change in the social or medical environment is necessary to allow these individuals to live as nearly as possible a normal life. Euthenics, to some degree, is accepted unconsciously as part of our lives. The wearing of glasses or contact lenses is applied medical euthenics. Printing books in braille for the blind, or producing television programs where sign language is substituted for or added to the audio portion of the show for the deaf are examples of social euthenics. So are the construction of ramps in public buildings, and the lowering of floor buttons within elevators. Making insulin available for diabetics and performing operations to correct deformities detected at birth are examples of applied euthenics. One of the greatest triumphs of euthenics is the correction of Rh incompatibilities.

In fact euthenics is the law of the land. In 1973 Congress passed a sweeping Vocational Rehabilitation Act for the mentally and physically disabled that stated: "No handicapped individual shall be excluded from any program or activity receiving federal financial assistance." The Education for All Handicapped Children Act that grants all disabled children the right to a free public education was passed in

1975. In April of 1977 the Secretary of Health, Education and Welfare signed regulations enforcing the legislation. Since there are few institutions in the country that are not receiving federal financial help, it means that provisions must be made in every school, public building, transportation system and practically every other facility used in our daily lives to accommodate the nation's 35 million handicapped. The mentally handicapped will no longer be in special schools, but will be part of the normal classroom routine. Buses will have doors wide enough to allow the entrance of a wheelchair. Provisions must be made for deaf and blind children in school classes with non-handicapped children. This is, indeed, euthenics with a vengeance.

It no longer seems that we must consider either a moral twilight in which the weak and deformed must suffer, or a genetic twilight in which deformities will continue to be passed on, as the only alternatives facing us. In fact, it no longer appears that we must consider either one of these as an ultimate probability. I consider the genetic twilight to be the lesser likelihood of the two alternatives for several reasons. First, even with no change in our approach to the problem a significant change in the incidence of most diseases with a genetic component would not take place for generations, in some cases hundreds of years. Combined with this slow evolvement is now an ever-increasing acknowledgment of the problem, and the responsibility of the individual to help solve it. Finally, there is the increasing knowledge of the genetic makeup of each individual, the consequences of producing offspring of those couples whose gene patterns are incompatible, and the future hope of being able to change gene composition.

It is encouraging to see the emphasis now placed on euthenics. However, on the other side of the question, there are still loud and persuasive voices raised with the message that the necessity to practice euthenics, the necessity to take care of those physically and mentally handicapped, is an imposition on society, and is caused by a situation that is easily corrected by simply preventing the handicapped from being born or eliminating them after they are born. Much of the rhetoric promoting the upgrading of the gene pool is based on the "right of the individual to be born with a sound physical and mental constitution based on a sound genotype." The motive certainly sounds admirable—to assure a healthy life for each individual. However, every such judgment presumes a knowledge of what is good or bad for the individual. I am suspicious that, in most cases, the underlying motive is to relieve society, rather than the individual, of a burden.

When we say genotype we mean the entire combination of genes present in the individual. However, now there seems to be a confusion between a sound genotype, those things affected by invisible genes, and the *pheno*type, the actual external appearance of the individuals. Through television and other media we have been conditioned to accept certain types of individuals, both male and female, as ideal specimens of the human race. There is probably nothing radically wrong with the idea that a certain physical appearance is a desirable characteristic of an individual. Neither is there anything wrong with the idea that it is desirable to have a perfect gene composition in an individual. What is wrong is the idea that anyone without a sound genotype—the combination of genes, or a sound phenotype—an attractive personal appearance, should not be considered a full member of society.

A sound genotype and an attractive phenotype—what do they mean? They mean nothing, as long as they are not used as standards to determine life and death.

So we have come full circle in considering eugenics as a procedure either for improving the race or improving individuals. Improvement is certainly a desirable end. Within eugenics there are both acceptable and unacceptable means for attempting this. The endless discussion goes on. What are the standards of a sound genotype? And who decides?

7 . . . *wherein control of reproduction and the development of new reproductive methods become realities*

Paralleling discoveries in genetic manipulation—even outdistancing these discoveries—have been the advances in what might be called reproductive engineering; that is, the modification of normal methods of reproduction.

It is hard to imagine a subject that, in view of our western religious and cultural tradition, might cause more apprehension than the suggestion that we tinker with the natural method of reproduction. But techniques have been developed for almost every phase of the reproductive process to the place where the need for a man and woman can almost be eliminated except as the manufacturers of a sperm and an egg. Cloning would eliminate even that requirement.

It is accepted as a fact that fertilization of an ovum may occur outside the human body. It is no longer only within the realm of science fiction to think of a fertilized ovum being transferred from a test tube to a human uterus. The knowledge of what sex an offspring will be is already available through amniocentesis, and it is possible to insure the birth of either a male or female child. Artificial insemination is a routine procedure. Only cloning, the asexual reproduction of genetically identical individuals from single cells, has not yet been accomplished in human beings, but it has been successful in plants and in frogs.

The first modification, and one that has been accepted clinically, is that of artificial insemination. It is not generally recognized how old the technique of artificial insemination is. The first recorded experiment in this field was in Pavia, Italy, in 1776, when the Abbé Lazzaro Spallanzani, a Catholic priest-biologist, tied "little breeches of oilskin" onto male frogs, collected the semen, and fertilized eggs with it. The same technique was used on dogs. Two early records of successful artificial insemination in women were recorded, one in 1799 by Home, and the other in 1884 by Pancoast. Although spermatozoa were originally seen under the first microscope of van Leeuwenhoek, it was not until 1856 that Pringsheim, a German biologist, actually observed a spermatozoid enter the egg shell of an algae, and proved the function of the sperm. Before that time even people such as Abbé Spallanzani thought that it was the seminal fluid that effected conception.

Artificial insemination of women is an established procedure today. While it is difficult to obtain any exact figures, it is estimated that there are probably as many as ten thousand successful artificial inseminations taking place each year in the United States. The acceptance of this procedure has reached the point where human sperm banks have been established in various parts of the country. It has been known for years that animal sperm can be preserved over long periods of time by freezing. The entire industry of animal artificial insemination is based on this fact. Semen can be stored under liquid nitrogen for at least ten years. In 1974 over 7 million cattle (5.5 million dairy, 1.5 million beef) were bred in this way, each championship bull being responsible for siring over forty thousand offspring in a single year. Human sperm can also be stored, with successful inseminations having been reported after ten years of sperm storage.

Both Edward Tyler of the Tyler Clinic in Los Angeles, and J. K. Sherman, of the University of Arkansas School of Medicine, have reported on successful inseminations with sperm frozen over long periods of time.[1] The percentage of birth defects in both these studies was below the average seen under normal circumstances in the general population.

In a survey of clinics using frozen semen, Sherman reported the same low incidence of defects occurring in 564 live births.[2] One of these births, that of a normal seven pound girl, is particularly noteworthy, since it involved the storage of semen frozen for at least six years and then shipped from the United States to South America where the successful insemination took place.

At the 145th meeting of the American Association for the Advancement of Science held in Houston in January of 1979 Dr. Armand Karow, Jr., of the Medical College of Georgia, reported that the use of frozen semen was rapidly increasing, involving about one in five inseminations. An unsuspected advantage of the use of frozen semen could be a reduction in birth defects and miscarriages. A survey conducted by Dr. Sherman of more than 1,000 children born as a result of frozen donor semen showed that less than 1 percent had birth defects, compared with 6 percent in the general population. Women who had become pregnant from frozen semen had fewer than 6 percent miscarriages, as against a rate of 10 to 15 percent in the general population.

Dr. Karow speculated that the freezing destroys defective sperm, thereby increasing the chance of only healthy sperm being transmitted.

Although most of the inseminations now occur with fresh sperm from the donor, several reasons have been given for the desirability of sperm banks. A man undergoing a vasectomy might want to deposit sperm so that he can become a father at some future time. Individuals undergoing radiation therapy might want to eliminate the possibility of genetic damage in future offspring. To solve the commonest problem of male infertility—inability to produce enough sperm with any single ejaculation—several samples of sperm might be preserved and then concentrated, thus assuring a high sperm count.

In either natural insemination or artificial insemination there is no control over the sex of the child to be conceived and born. Since both male and female sperm are present, the chance of a male or a female being born is about 50-50. Sex determination is an objective of those interested in the total control of the reproductive process. When I talk about sex determination, I think of two extremes possible in accomplishing this end. The first is assuring the conception of either a male or a female, and the second is assuring the birth of either a male or a female. I use the word "determination" in this case not as meaning simply pre-knowledge.

If one is satisfied to accept the method of allowing only a male or a female to be born, the technique is already available. Using amniocentesis, the sex of an embryo can be known early in the period of pregnancy. If it is not the desired sex, an abortion is performed. When the desired sex is conceived, the child is allowed to be born. Admittedly, this is a clumsy, time consuming, unattractive, and, to me, immoral procedure. Nevertheless, it is available, and is being used.

To determine sex at the time of conception, present knowledge indicates that it is the sperm that must be manipulated to be sure that one or the other of the sexes is being born. The sex of the child is totally dependent on whether the ovum is fertilized by either a male or a female sperm, both of which are present in semen.

One of the obvious ways, then, to choose the sex of the child would be to inseminate artificially with semen separated into male and female components. To accomplish this it is necessary to develop a method to assure that only preselected male sperm or female sperm will have a chance to fertilize the egg. The X-chromosome sperm, the female sperm, is more dense, having about three percent more genetic material than the male Y-chromosome sperm, and so possibly could be separated by some method involving density, that is, centrifugation or settling. Indications are that the sperm move or swim at different rates of speed. So far, all attempts at separation, in both the animal and human field, have met with failure. Other approaches have involved attempts to take advantage of chemical and biological differences, and so destroy or inactivate one sperm, while leaving the other. Even vaccines against one or the other have been proposed. However, the differences between the structures are so small that no separation has yet been made.

Through the ages various procedures have been recommended for influencing the selected birth of a boy or a girl. Aristotle, for example, suggested that having intercourse in the north wind resulted in male children being born, while the south wind was a more advantageous environment for the conception of females. In more recent times it has been suggested that controlling the movement of sperm by controlling the acidity or basicity of the vagina would allow the male or female sperm to move faster, and thus have a better chance of being the first to fertilize the ovum. Similarly, it has been suggested that acid conditions harm the male sperm more than the female. However, none of these methods has been proved to be effective. Modern procedures seem to be no more successful than the one proposed by Aristotle.

The stimulus for the preselection of the sex of a child comes almost totally from the parents, usually based on emotional reasons. The desire to have a male as the eldest, the desire to have a child of each sex, the desire to have a male to carry on the family name, or just the fact that the parents may prefer one sex over the other is usually the only reason for pre-sex selection.

Much of the public concern over such sex-determination techniques arises from the impact of the possible changing of the present sex ratio. Presently, 51.45 percent of babies born are males. The ratio becomes equal later in life because of the shorter life span of the male. Data on which to base a prediction of the consequences of sex selection are hard to come by, representing as they do the determining of so much subjective opinion. The best study seems to be that of Westoff and Rindfuss of the Office of Population Research at Princeton. They concluded that the principal effect in the United States would be a significant increase in the number of first-born who were males, with the second child being a female. They further concluded that there would be an initial and temporary surplus of males in the first couple of years, but that this would eventually be equalized by a balancing number of female births. In one-child families, both here and in other countries, the number of males would be expected to increase.[3]

Aside from the purely subjective desire to have either a male or a female child, a positive benefit from being able to determine the sex at conception would be the possibility of preventing the birth of babies with sex-linked genetic disorders. For example, hemophilia is a genetic trait carried by females, but the disease is passed on only to males. Therefore, if no male children are born to a hemophilia-carrying mother, there would be no clinical hemophilia possible. However, if the mother bore only girls, the result would be an increase in the number of hemophilia female carriers (depending on how many inherited the gene from the mother).

A much more controversial aspect of reproductive engineering lies in the so-called test tube babies, and the use of surrogate mothers using eggs fertilized either *in vitro* or naturally in a mother's womb. In addition to the obvious ethical problems, there are still unsolved and even untried legal issues involved.

In normal human reproduction, fertilization of the female egg occurs when it is penetrated by a sperm. The cell divides into two cells, then into four, eight, sixteen, and so on until, after thousands of divisions, a human body is formed. The initial division occurs as the egg passes down the fallopian tubes. After about three days the fertilized egg has reached the sixteen-cell stage, called a "blastocyst," and has passed from the fallopian tube into the uterus, where it remains for another three or four days, continuing to divide. At this stage, it either passes through the uterus and is eliminated, or it attaches to the wall of the uterus and continues to develop. The time at which it attaches to the uterus is termed "implantation," and it is from this time that symptoms of pregnancy begin.

The *in vitro* fertilization of animal ova is not new. The ova of hamsters, rabbits, mice, guinea pigs, rats, cats, gerbils, pigs, and cows have all been successfully fertilized. The period of gestation for a mouse is twenty-one days. Implantation of the blastocyst occurs after five days. Y-Chih Hsu, of Johns Hopkins, has succeeded in culturing mouse embryos from the blastocyst stage to a stage equivalent to that seen after nine days of gestation *in vivo* (inside the womb). This would correspond roughly to a four-month-old human embryo. The mouse culture is thus in a period well past the time when implantation occurs *in vivo*.

In addition to the study of *in vitro* fertilization, experimenters have attempted to transplant fertilized eggs of animals into the uteri of other animals and observe the normal growth. This technique has been developed to the point where it is now almost routine. Some years ago an experiment in Australia received considerable publicity. A fertilized egg was isolated from the uterus of a ewe and transplanted to the uterus of a rabbit. The rabbit was flown to England, where the growing egg was again transplanted to the uterus of a sheep, and a normal lamb was eventually born.

Much work in this area was stimulated by the recognition of the beneficial effects that some of these procedures would have on agriculture. As early as 1950 a total concept of control from *in vitro* fertilization to growth in an artificial womb was proposed for the growth of cattle.

It is known that certain steroidal products will cause some animals, including humans, to produce more than one egg during an ovulation cycle. These products are called superovulators. If such a superovulator is administered to a cow, a number of eggs can be isolated at any one time. The eggs can be fertilized *in vitro,* using bull semen. Semen is composed of male and female sperm. Since the sex sperms are different, and on the assumption that any two things that are different are separable, the egg would be fertilized with either male or female sperm, thus determining sex of the offspring. Each fertilized egg would then be transferred to the uterus of a cow that had been prepared to receive it, and develop into a normally-delivered calf. The reason for desiring this procedure is obvious. Since the genetic characteristics of the animal are determined by the sperm and the egg and not by the animal that is used as the carrier of the fetus, it would be possible to get numerous, maybe hundreds, of offspring each year from each genetically superior cow and bull, instead of being limited to only one if the original cow must carry the calf to term. Any scrub cow can be used as a carrier with only proper nutrition being required. The ultimate will be the development of an artificial uterus, eliminating even the necessity for a carrier. With the exception of separation of male and female sperm, and the use of an artificial uterus, this cycle is now practical. In fact, there are commercial organizations now making this service available to cattle producers. Other animals whose embryos have been transferred and successfully implanted, although the embryos were not necessarily previously maintained in culture, include mice, rats, rabbits, pigs, sheep, and horses.

In June of 1976 scientists at Texas A&M, the University of Texas, and the Southwest Foundation for Research and Education reported that a full term, normal infant male baboon had been born on September 5, 1975 as the result of the transplantation of a five-day-old embryo from its genetic mother to another female baboon. The fertilized ovum had resulted from the natural mating of the original mother.

The first report of an *in vitro* fertilization of a human ovum by a sperm, that is, fertilization outside the womb, was reported in 1950 from Dr. Landrum Shettles of Columbia University. In 1961, Dr. Daniele Petrucci of the University of Bologna reported a fertilized egg had lived *in vitro* for 29 days. However, neither of these reports is well documented scientifically. In the last few years reports in the scientific literature of Dr. Robert G. Edwards and his group at Cambridge University leave no doubt that in vitro fertilization had indeed taken place. Speculation began as to whether or not the next step—the transplantation to a human uterus—would be undertaken.

Then, on July 25, 1978, came the dramatic announcement that a normal child had been born, conceived *in vitro* from her mother's ovum and her father's sperm. After twelve years of work the team of Doctors Steptoe and Edwards had finally made real what had been previously science fiction.

The mother, Mrs. Lesley Brown of Oldham, England, could not bear children because of blocked fallopian tubes: an egg could not travel down the tube and be fertilized in the process. To overcome this difficulty Mrs. Brown was treated with a hormone to stimulate ovulation, and then an egg was removed surgically from one of her ovaries. The egg was placed in a dish (not a "test tube") containing a special nutrient to keep it alive. There it was bathed in the husband's sperm. When one of the millions of sperm present penetrated the egg, fertilization had taken place and cell division began. After two and a half days the egg was transferred to Mrs. Brown's womb, which had been prepared for pregnancy by the administration of other hormones. A normal term of pregnancy resulted, and Louise Brown was born.

Reaction to the announcement of the birth ranged from elation to total condemnation. Some scientists welcomed it as a "stunning achievement," others were apprehensive about the excessive enthusiasm generated by the birth of only one baby, and cautioned that the procedure was still experimental and the possibility of genetic damage and subsequent birth of subnormal children was still a possibility.

The publicity attending the birth of Louise Brown attained proportions usually thought to be reserved for the report of the Second Coming. Worldwide press, radio, and television services competed for the most minute bit of news associated with the new baby, her parents, doctors, or acquaintances. In a Louis Harris Survey published in *Parents Magazine* shortly after the birth, 85 percent of American women reported their belief that couples who can't have children should be able to resort to test tube fertilization. More than half (58 percent) of the women of childbearing age in the survey would consider using the method themselves.

On January 14, 1979, the second successful birth, this time of a male baby in Glasgow, Scotland, was reported by Drs. Steptoe and Edwards. Of the two major Chicago newspapers, one carried an item about the birth on page 10, the other ignored the story completely. In less than six months an event originally considered almost miraculous has now come to be considered almost normal.

This is a dramatic example of the almost passive acceptance of a revolutionary advance of science once the advance has taken place. The birth of one or two or a dozen babies as a result of embryo transplants solves none of the ethical, moral, or even medical problems that existed before the first birth. It only proves that the process can be carried out successfully. The problems of safety, of predicted unsuccessful transplants and the resulting destruction of the fertilized eggs remain. But science has been successful, as science said it would be. So embryo transplant is now an accepted fact and, for better or worse, will probably also be an accepted and acceptable procedure.

The hope that this brings to infertile couples has been highly publicized, possibly too highly publicized. Certainly all infertile couples would not be candidates. The experimental character of this procedure and the risk of genetic damage to the child cannot be overestimated. And finally, the procedure is very expensive and must be carried out only by experts under the most carefully controlled conditions.

The arrival of the first *in vitro* baby did not result in any new moral or ethical problems. Rather, it simply transferred their consideration from the realm of the theoretical to the real. The greatest objection was to the fact that, in order to obtain a successful transplant, several eggs would be fertilized, and then only one would be selected for implantation. The others would be destroyed. If you believe that a fertilized egg is human from the moment of fertilization this would obviously be a critical moral problem. Possibly because all of the arguments for and against *in vitro* fertilization had been presented vigorously while the subject was still hypothetical, the reaction to the actuality was surprisingly light from church leaders opposed to the procedure. The arguments against the unnatural process divorcing intercourse from procreation were reiterated, as was the stand of the Catholic church against artificial insemination. But now that an actual *in vitro* baby was being considered, the moderate views expressed by religious leaders from whom vigorous opposition might have been expected were surprising. Bishop Cornelius Lucy of Cork, Ireland, said, "Offhand, I don't see anything wrong with childless couples using test tube methods if there is no other possible way for them to have babies." Bishop Augustine Harris of Liverpool said that he would tell a couple to "go ahead" if they had decided they desire the child because of love for the child and for each other, and if they believed it was within the context of their marriage. Father Bernhard Häring, a prominent Catholic theologian, Bishop Mark Hurley, Chairman of the United States Bishops' Human Values Committee, and Cardinal Gordon Gray of St. Andrews of Edinburgh, Scotland, all expressed the opinion that more study is necessary before moral judgments can be made.

At the exact time that Louise Brown was becoming a part of

history in England, another kind of related history was being made in New York. In 1973 John and Doris Del Zio had decided that they, as a last chance at producing a child of their own, would seek assistance in having an *in vitro* fertilized ovum transplanted into Mrs. Del Zio's uterus. Dr. Landrum Shettles of Columbia Presbyterian Hospital had been reported earlier to have fertilized ova *in vitro,* but had never attempted a transplant. It was to him that the Del Zios went for help. On September 12, 1973, an egg was removed from Mrs. Del Zio by surgery. The egg was given to Shettles who proceeded with the fertilization, using the husband's sperm. But the Del Zios were not to be the parents of the first *in vitro* fertilized child.

When the Chief of Obstetrics and Gynecology heard of the experiment he decided to terminate it because of the possible birth of what he termed a "monstrosity," and because the experiment had not been cleared by the hospital's Human Experimentation Review Board. In fact, he exposed the contents of the flask containing the egg and sperm to air, thus rendering the culture useless.

The Del Zios sued the Chief of Obstetrics, Columbia University, and Columbia Presbyterian Hospital for $1,500,000. The trial started on July 17, 1978, the week before Louise Brown was scheduled to be born. On August 18 a jury awarded the Del Zios $50,003 in damages on the grounds that psychological harm had been inflicted on the couple. The husband was awarded one dollar from each of the three defendants. However, the jury found the defendants innocent of wrongfully depriving the Del Zios of property.

The reality of *in vitro* fertilization and surrogate mothers raises again the problem of legal parenthood. Who has the legal right to a baby born as the result of an implanted fertilized egg into a surrogate uterus? Possibly the legal position is more easily defined when the wife's egg is fertilized with the husband's sperm. It would seem clear that the husband and wife were the true legal parents.

However, the situation caused by artificial insemination raises some legal doubts. If a husband and wife, unable to bear children because of some deficiency in the husband, agree to have a child by artificial insemination using donor sperm, does the donor have a legal claim to parental rights? In England, a husband and wife, unable to bear children because of a physical difficulty in the wife, decided to use a surrogate mother artificially inseminated with the husband's sperm. When the baby was born, the mother, a reformed prostitute, decided she wanted to keep the child in spite of the agreement she had made with the childless couple. When the case came to trial, the English judge awarded custody to the woman who had borne the child.

Several reasons are given in order to rationalize and justify such a radical departure from generally accepted social and ethical norms. First, it could be the answer to those infertile couples who want children of their own. For example, a woman who has blocked fallopian tubes cannot become pregnant. However, it is possible to remove an ovum, have it fertilized by her husband's sperm and retransplanted into her uterus, thus by-passing the tube. Certain genetic characteristics such as sex can be determined by an examination of the blastocyst. Thus, only the blastocyst with no identifiable abnormalities would be transplanted. Some women are a definite health risk in carrying a fetus to term and giving birth. Surrogate mothers could be used to produce a natural child of parents who otherwise would not be able to pass along their own flesh and blood. Finally, I suppose there will occur the time when surrogate mothers will be hired simply to relieve other women of the inconvenience caused by a nine-month pregnancy.

Work on an artificial womb for human beings has been in progress for some time. Originally, this work was started in order to preserve the lives of embryos aborted in very early stages of development. Dr. Robert Goodlin at Stanford University was one of the early workers in this field. Drs. Theodore Kolobow and Warren Zapol at the National Heart Institute developed a womb that has kept lamb fetuses alive for several days.

Based on what is known from animal studies and what is known about the stability of abnormal embryos in human beings, the risk of producing abnormal offspring is probably no greater, if even as great, than would result from normal reproduction. As has been mentioned, the *in vitro* fertilization and implantation of ova into animals is a routine procedure. In these cases there is no increase but, in some instances, a decrease in abnormal young. In humans, a very high percentage of chromosomally abnormal embryos spontaneously abort very early in pregnancy. Therefore, it could be assumed that, should abnormalities result from *in vitro* manipulation, the resulting fertilized egg would not develop to birth.

What is probably the most revolutionary development in the

control of reproduction in the twentieth century came with the introduction of the oral contraceptives in 1960. For the first time it was possible to prevent conception with practically 100 percent certainty by a convenient and simple method. The clinical development of birth control pills is a fascinating example of how research results obtained in two programs with diametrically opposite objectives were combined to solve both problems.

Dr. John Rock, Harvard professor of gynecology, was director of the Free Hospital for Women in Brookline, Massachusetts, working to solve the problem of infertility in women. As the result of some of his past work he had concluded that at least some of the women might be cured if he could cause a pseudo-pregnancy, thus bringing about a change in the physiology of a woman and stimulating those systems that, because of their improper functioning, had prevented pregnancy. When treatment was stopped, if the theory was correct, the physiological system should react normally, in other words, rebound, and pregnancy would be possible. For this purpose he had been using large doses of the synthetic estrogen diethylstilbesterol, combined with another hormone, progesterone. The pseudo-pregnancies, with all their undesirable symptoms, did occur. Following cessation of treatment a large number of women, previously barren, did become pregnant. However, because of the undesired side effects of the treatment, Dr. Rock was not satisfied, even though his final results were good.

While this work was going on, Dr. Gregory Pincus, an old friend of Dr. Rock, was carrying out work only 40 miles away at the Worcester Research Institute, work aimed at preventing pregnancy. Working with mice he had found that progesterone alone, one of the compounds being used by Dr. Rock, inhibited ovulation and thus prevented conception. When Pincus and Rock exchanged research results Rock decided to use progesterone alone to see if he could get the results obtained previously with his combination. At the suggestion of Pincus, he also administered drugs only 20 days during each month instead of daily. During the trial, studies showed that almost all the women had ceased to ovulate, yet there were none of the effects of the pseudo-pregnancies reported earlier. When the women were taken off the treatment, encouraging results were obtained. A number of women became pregnant. The effect now known as the "Rock Rebound" had been the same.

However, the product did not seem to be completely satisfactory from the points of view of either Pincus or Rock. The women studied by Rock showed some break-through bleeding, and, a disadvantage for Pincus, the progesterone was only 85 percent effective in inhibiting ovulation.

At this point the G. D. Searle Laboratories, a pioneer in steroid hormone research, submitted to Dr. Pincus for tests some synthetic chemical compounds that, while having the effects of progesterone, also acted differently in some respects, and also were more active when taken by mouth. Pincus tested a large number of these products, as well as those from other laboratories. Two of the Searle products he found to be many times more active than progesterone. Rock then tested the Searle products and found them to be active in his patients, also, but at lower doses and with fewer side effects than had been observed in his earlier patients.

The first oral contraceptive was introduced in 1960, a little over ten years after Pincus started his first work in mice. Thus it was that the same product worked on by two different research groups was found to be the answer to two different problems, preventing pregnancy *and* making pregnancy possible.

It should be mentioned that the fact that the oral contraceptives were made available in 1960 was as much the result of a business decision as it was of scientific research. Previously very little research had been conducted on contraception, particularly in commercial laboratories. The reason had very little to do with either the ethics or morals of contraception. It was simply that the companies did not want to face what they assumed would be strong criticism from the Catholic church. Had it not been for this, it is very probable that some product would have been introduced for this purpose years earlier. In fact, the product that was eventually approved in 1960 for contraception had been on the market for three years before as a treatment for threatened abortion and menstrual disorders. The efficacy of the product as an effective contraceptive had been whispered around, but no official claims for that purpose were made until 1960.

When disaster did not strike Searle, a number of other firms quickly began work and, within a few years, other products were available. Some of them are made up of chemical compounds that had been made years before but had never been tested because of the reluctance to enter this field.

Since the introduction of the first oral contraceptive, only products to be taken by the woman have been introduced. This, contrary to some opinions, is not due to male chauvinist chemists. It is simply easier to interfere with the more sensitive reproductive system of a woman than of a man. There has been much research directed toward finding a male contraceptive. So far nothing practical has resulted.

The same society that recognizes the importance of controlling population growth also seems, at the same time, to recognize the rights of parents to have children, and to approve of efforts to restore fertility to childless couples. The medical profession and the research community have recognized for years that a method of allowing a hitherto infertile woman to bear a child would be a true medical advance, an advance in the real medical sense that it would relieve human suffering. The unfulfilled yearning for a child of one's own flesh and blood is, indeed, as much a suffering as are many of the other disorders afflicting people. Society approves of the use of the so-called "superovulators," fertility drugs that stimulate ovulation. The side effects of these drugs are well-known. Multiple births are quite common, with four or six embryos resulting in some cases, and as many as twelve having been reported. Practically all of these fail to survive.

About 5 percent of all the women of childbearing age in this country are infertile because their fallopian tubes are blocked. Thus, although ovulation takes place, the egg cannot pass through the tube or be fertilized.

I must admit it is repugnant to think of a laboratory whose only function is to be a baby factory. It is hard to imagine having as an objective only the making available of fertilized eggs simply to produce more of a particular kind of human being. It is not quite so hard to imagine, but it is equally objectionable, to think of a surrogate mother being used only to relieve discomfort and inconvenience for the biological mother.

In vitro fertilization of human ova could certainly be thought of as a dehumanizing act. It is dehumanizing if its objective is to eliminate the natural, loving union of a man and a woman that results in the birth of a child. However, it is hard to imagine anything dehumanizing about a wife's ovum fertilized by her husband's sperm being implanted into the wife's uterus to undergo the same gestation period as would an embryo conceived under normal conditions, and resulting in a child, possibly more wanted and more loved because extraordinary means had been used to bring it into being.

The ultimate in genetic engineering is "cloning," a popular subject for speculation these days. Cloning is the term applied to the asexual reproduction of exact duplicates of individuals. It involves transplanting whole nuclei that carry the DNA from one cell to another. Cloning has already been carried out in plants, such as carrots, and in frogs. Dr. J. B. Gurdon of Oxford University, working with African clawed frogs, took an unfertilized egg, destroyed its nucleus with radiation, and replaced it with the nucleus of a cell from the intestine of a tadpole. The result was an exact genetic duplicate, not of the frog from which the egg had been taken, but of the tadpole.

It is known that in humans every differentiated somatic cell contains in its nucleus the same complete set of genes and, therefore, all genetic characteristics of the individual. It could therefore be postulated that any skin cell, for example, could be used to start egg division, the egg transplanted to a uterus, and exact copies of the donor produced.

As a result of the development of this technique, Orwellian predictions of its eventual use have become a favorite diversion, not only of science fiction writers, but of science journalists. Since the people resulting from cloning procedures would be exact duplicates of the individual donating the cells that are transplanted, any desired characteristics could be assured. Further, they could be produced in any number. Even further, the selection would not have to be made until all the characteristics of the donor had been determined through his life. Scientific or literary geniuses could be duplicated. Beautiful women and handsome men—whatever these adjectives mean—could be produced in quantity. One can even conjure up a vision of identical seven foot, All-American basketball players, or giant football players being developed. Whether the alumni would be willing to wait eighteen years for such a development is debatable. It has even been proposed that unscrupulous dictators might use this method to develop thousands of pliable, uncomplaining, unquestioning slaves to do their bidding. While it is an interesting science fiction concept, one of the factors mitigating against such an occurrence is the time involved. The individuals would have to go through the same development, from birth to adulthood, as does every other human being. Therefore, from the time individuals are cloned until they come to an age when they could be "used" would probably be about sixteen or eighteen years.

In 1978 a book was published purporting to be a record of the first cloning of a human being. According to the story, a west coast millionaire, wishing to leave posterity a clone of himself, approached a science writer to see if arrangements could be made to have such a procedure undertaken. The writer "after a long period of soul searching" recruited the scientific talent, and the experiment was performed outside the United States in a land "beyond Hawaii." At the time of the announcement of the book, the resulting clone was supposed to have been fourteen months old. The report says that ova were collected from women who believed they were helping infertile women bear children. The nucleus of each egg was removed, and was replaced with the nucleus of a testicular cell from the millionaire donor. An egg began to divide, and was implanted into the uterus of a surrogate mother. The birth allegedly took place normally in 1976.

Scientists generally discount the story. Certainly, cloning is theoretically possible. However, techniques, even in the most advanced laboratories, have not approached the stage where it yet appears practical. It appears unlikely, then, that qualified scientists could have been recruited to perform such an operation successfully. No scientific evidence is presented to back up the claim. If the donor and clone do exist there are at least fifty tests available to substantiate the clone. Technically and genetically the donor and the clone would be identical twins, rather than father and son.

In addition to the time involved in developing a clone there are two other practical objections to cloning. The first is that, regardless of the identity of genetic characteristics, individuals would probably react differently in different environments, and the second is the undesirability of narrowing the variation of genes in the so-called "gene pool." It is now generally agreed that characteristics of individuals are determined by the interaction of genes and the environment. As has been pointed out this is one of the fallacies in the belief that cloning will produce individuals who will act exactly the same as the original source of the clone. They will act identically only if they have been exposed to the identical environment.

Much has been made of the fact that individuals carry different genetic variants, some of them lethal under the right (or wrong) circumstances. Yet there is a body of genetics that says species survival may depend on keeping a large number of these variants in the "gene pool." If everyone had exactly the same gene makeup everyone would also suffer from the same weaknesses as well as strengths. It is only where there is wide variability that we can be assured of survival and adaptation to the stress of environmental change. Agricultural geneticists can give the basic answer to why genetic variation is desirable. They call it "hybrid vigor."

It is this same principle that is the basis for the scientific objection to the eugenic procedure of selecting a number of "supermen" whose sperm would be used to inseminate artificially a large number of women, thus, presumably, upgrading the level of the race to that of the donors. Aside from the fact that the same characteristics might not be duplicated in a different environment, all the individuals would have the same inherited physical weaknesses. A change in environment that would emphasize a weakness could thus eliminate a large part of a population, whereas a large part of a population with differing characteristics would adapt and survive. It is thus that evolution takes place.

It is impossible, by definition, for everyone to be heterozygous for any one gene. As soon as everyone living in the world became heterozygous the first offspring from a couple would be homozygous. The value lies in many people being heterozygous for different genes. Thus, the paradox. A gene pool is necessary for survival. At the same time clinical genetic defects are caused by the chance union of two people with the same deleterious recessive gene. The idea that genetic diversity is good has encountered much antagonism. If we say that the greater the number of genes there are in the human gene pool the greater will be our ability to adapt to changing environments, it is interpreted by some as a pulling back from a position that says that every effort should be made to eliminate abnormal genes from the race. This fear, in turn, is based on the belief that this is one way to eliminate educational failure, crime and poverty. Just the opposite is true. Genetic diversity is a probable necessity for the development of the human race, and the recognition and acceptance of diversity is a much surer way to approach social equality.

It is a generally acknowledged principle that what can be done will be done. *In vitro* fertilization and transplantation of the fertilized ovum, gene manipulation, even cloning *can* be done. However, I have a deep and abiding faith in the human instincts that lead me to believe that these techniques will not be misused. The very things that make us human will force us to reject whatever decreases that humanness.

8 . . . *wherein composite human beings are made and their minds controlled*

Modifications in the human organism brought about by gene changes are based on intervening in beautifully subtle, exquisitely sensitive mechanisms that depend on hundreds of thousands of internally controlled interactions. At the other extreme there are procedures that can bring about equally radical changes in the human body and even the human mind by the direct intervention of operations on specific organs, replacement of organs, or the external stimulation of organs.

The fact that such procedures seem straightforward, almost clumsy, compared to the complexities involved in genetic changes, does not make the ethical problems involved any less serious. The one factor that does make the consideration of these procedures more acceptable is the fact that results are limited to the individual on whom the procedure is practiced. Changes are not passed on to offspring. Nevertheless, even though the procedures involved are closer to those things to which we have become accustomed, procedures performed by surgeons in operating rooms or shock therapy performed for resuscitation or relief of mental diseases, they involve the same moral, ethical, and legal problems that we associate with other forms of biomedical engineering.

It seems like a very long time ago, and the debate seems almost trivial in view of the problems facing us today, but the ethics of organ transplants were once questioned with almost the same intensity as the morality of abortion. While the debate still continues around the ethics of removing an organ from an individual being kept alive by artificial means, the objection to organ transplants themselves on ethical grounds has almost entirely disappeared.

I suppose the precursor to organ transplants was blood transfusion. From earliest times blood was thought to be the basis of life.

The ancient regard for life is reflected in the Book of Leviticus: "Because the life of the creature is in the blood, and I appoint you to make expiation on the altar for yourselves: It is the blood, that is the life, that makes expiation."

In Greek mythology when Jason wanted to have health and youth restored to his father, Aeson, he called on Medea to use her magical powers to purify the blood of Aeson. She cut the old man's throat so that all his blood poured out. This she replaced by a potion, and, as a result,

> his beard and hair lost their hoary gray and quickly became black again; the pallor disappeared and look of neglect; the deep wrinkles were filled out with flesh; and his limbs had the strength of youth. Aeson was filled with wonder, and remembered that such he had been forty years before.

Would that it were true!

The first real study of transfusions from animal to animal was carried out in 1665 at Oxford University by Richard Lower. This was quickly followed within a year by transfusions of animal blood to humans.

One of the early records of these transfusions is contained in Samuel Pepys' Diary. It may lack the scientific authenticity of a *Lancet* or *Journal of the American Medical Association* report, but it makes up in human interest what it lacks in experimental detail:

> November 14, 1666. . . . Dr. Croone told me that, at the meeting at Gresham College tonight . . . there was a pretty experiment of the blood of one dog let out, until he died, into the body of another on one side, while all his own ran out the other side. The first died upon the place, and the other very well, and likely to do well. This did give occasion to many pretty wishes, as of the blood of a Quaker to be let into an Archbishop, and such like; but as Dr. Croone says, may, if it takes, be of mighty use to man's health, for the amending of bad blood by borrowing from a better body.

> November 21, 1667. With Creed to a tavern, where Dean Wilkins and others; and good discourse; among the rest, of a man that is a little frantic . . . that the College have hired for 20s. to have some of the blood of a sheep let into his body; and it is . . . about twelve ounces; which, they compute, is what will be let in in a minute's time by a watch. . . .

> November 30, 1667. . . . I was pleased to see the person who had his blood taken out. He speaks well, and did this day give the Society a relation thereof in Latin, saying that he finds himself much better since, and as a new man; but he is cracked a little in his head, though he speaks . . . very well. He had but 20s. for his suffering it, and is to have the same again tried upon him: the first sound man that ever had it tried on him in England, and but one that we hear of in France. . . .

Transfusion has become such a routine part of human medicine today that it is hard to accept the fact that not too many years ago there was considerable opposition to this procedure on ethical grounds. In particular, a concept of different human characteristics being associated with the blood of different races persisted until relatively recent years. We now know that the differences are immunological, expressed as blood types, and that except for the inclusion of viruses or bacteria, blood transfusions from one individual to another individual of the same type are safe and effective, and have no effect on the physical or mental characteristics of the recipient.

With the advent of organ transplants questions arose again as to the ethics of using the parts of one person's body to serve another, either before or after the death of the donor. From the time of the alleged transplant of a human heart into General Kuan Kung by the famous Chinese surgeon Hua T'o who lived in the third century A.D., philosophers and science fictionists have speculated on the potential and ramifications of the extension of the exchange of body parts and mental processes from one being to another. It was not until transplants became a reality, however, that public concern became evident. The first heart transplant took place in December of 1967, and caused considerable concern. Fortunately, such concern was short-lived. Such extreme questions arose as to what happened to the individuals who received a human heart. Were they still their original selves, or did they become the individuals who were the donors? Such thinking reflected the almost superstitious belief that the heart is really the center of the person, that emotional reactions are determined by the heart, and that, therefore, a person's character and identity are determined by his or her heart. Thus, a "hard heart," a "soft heart," a "broken heart." These issues were never taken very seriously, and now transplants are a routine part of medicine.

One of the first heroic experiments in modern transplantation took place in 1948 when Dr. David Hume at Boston's Peter Bent Brigham Hospital attached the kidney from a dead person to the arteries and veins in the wrist of a young woman dying of kidney failure. The kidney worked perfectly and after three weeks it was removed. The woman's damaged kidney, in the meantime, had been given a chance to recover, and the woman departed the hospital cured. Since then, 30,000 kidney transplants have been performed around the world, obviously using a different technique. In 1977 there were about 4,500 kidney transplants in the United States alone.

About 5,000 cornea transplants will be performed in the United States using material collected at more than 80 eye banks. There were 29 liver transplants worldwide in 1976. Even lungs and pancreas are being transplanted. In some cases, kidneys and corneas for example, the problem is no longer perfecting the operation but obtaining the organs. There is an organized effort to convince individuals to will their bodies to medicine after death, so that their organs may be used for the benefit of the living. Even such a conservative organization as the Catholic church has approved this practice. It has even been suggested that "body banks," places where bodies whose functions are kept alive by resuscitators, heart stimulators and intravenous feeding, be established as a source of fresh parts.

The one remaining issue is the protection of the donor from premature removal of an organ for transplant. There seems to be total awareness of this problem in the medical community. The subject was discussed in Chapter 2.

Transplant of hearts, kidneys, and cornea have become a part of medicine. There are records of arms, legs, fingers and toes being successfully rejoined to the body after they have been severed. Blood of an individual can be removed totally and replaced by different blood, even in a fetus still in the womb. So far, there has been no suggestion that a human head or a brain be transplanted, although I well remember reports and pictures of such operations on dogs performed by Russian scientists a number of years ago.

Since, unfortunately, there is no known wisdom-enhancing gene, nor is there much possibility of transplanting a brain, the future of influencing intelligence by the kinds of procedures we are discussing seems to be remote. However, there are methods, both physical and chemical, that can be used to influence people's minds and to subjugate their wills.

When we talk of drugs that influence behavior, I believe most people today would think of drugs that affect the mind. However, there are many other ways to affect behavior. Eating habits are very characteristic of individuals. There are drugs that will either stimulate or inhibit the appetite. Sleep pattern is a characteristic of a person. There are drugs to induce sleep or prevent it. Drugs are available to relieve pain, to stimulate or inhibit sexual drive. Alcohol, the psychedelic drugs, and narcotics certainly influence behavior. However, these drugs are not my concern in this discussion. Rather, I am interested in those drugs or procedures that allow one individual to exercise control over the will of another with or without the controlled individual's consent. I will exclude the activities popularized in countless television plots where the narcotics pusher encourages the victim to become addicted and then forces him or her to act by threatening to withdraw drug supplies.

These are problems involved in the changing of a person's behavior, but the key concept is not change, it is control.

Behavior control has been defined as getting people to do someone else's bidding. Possibly a general discussion of the ethical aspects of behavior control would be in order before discussing specific procedures. These would include the practice of psychiatry, which certainly in many cases has the same result—the shaping of behavior—as does a drug or a physical procedure.

In any situation involving a proposed behavioral change the prime question is "Why is the change being effected?" If it is for the good of the patient many of the objections now being raised regarding such techniques are eliminated. However, sometimes we can be deceived into thinking it is for the good of the patient when actually it is for the good of society, friends, or relatives, and not necessarily the patient. Society feels more comfortable if individuals conform. Therefore, anything that makes people conform is good. Such is not always to the benefit of the individual. If an aggressive, abrasive individual is changed into a more mild-mannered, retiring person, he or she will no longer cause problems for others. However, the patient's entire success and self-satisfaction might have depended on that aggressiveness. Society is pleased—the individual is harmed.

In traditional psychotherapy, patients have a chance to understand the implications of their treatment because it usually happens over a period of time. They even have a chance to resist it, since psychotherapy involves bringing patients to an understanding of their symptoms, to an awareness of their environment, and to the significance of changing that environment. There is at least the element of informed consent if they understand these things. However, with other forms of behavior control, the same time element is not present, so that patients do not have the opportunity to weigh the consequences as treatment takes them further and further along.

The next step up the line from classical psychiatry is conditioned response, the training of an individual to respond automatically to an outside stimulus or to perform a certain action in order to receive a reward or punishment, always forthcoming as a result of the action.

I suppose if I had to select one device that has controlled our lives more than any other single thing it would be the clock. There are very few people who do not respond automatically to the dictates of the time of day. Even our nervous system responds. How often have you become hungry when you suddenly see a clock that tells you it is past mealtime? Or how often sleepy when it is past bedtime? The traffic light is another insidious controller of our actions—we get a thrill from speeding up and sneaking through on the yellow, or approaching the light at a speed that allows us to go through just as the light changes to green. Clocks and traffic lights are taken for granted in our lives, but they are examples of behavior control.

Finally, there are those procedures in which the brain itself is acted upon in some way other than by drugs. There are records going back to the early Romans showing that insanity had been cured by a sword wound in the head.

However, the modern practice of psycho-surgery dates only from about 1936, when Egas Moniz, a Portuguese neurologist, relieved uncontrollable psychotics by performing what has become known as a prefrontal lobotomy on them. This work was predated by experiments of James Fulton, who caused chimpanzees to endure laboratory stress tests with what Fulton called "philosophical calm" by cutting certain groups of nerve fibers from the frontal lobes of the brain. In 1949 Moniz was awarded the Nobel Prize for his work on human beings. It is ironic that Moniz later met a violent death at the hands of one of his crazed former patients.

The use of psycho-surgery, defined as the selective destruction of brain tissue to alter behavior, has been used increasingly in recent years. The National Commission for the Protection of Human Subjects of Behavioral and Bio-medical Research reported that about 500 psycho-surgical procedures were performed each year from 1971 to 1973 by about 140 neuro-surgeons. Concern for the patients exposed to this procedure is indicated by the fact that Congress singled it out as one of two techniques (fetal research was the other) for special consideration by the same commission.

The Commission issued its report in March of 1977. In view of the fears expressed by both the medical community and the public concerning abuses to which this operation might be put, the report was surprisingly favorable. The Commission found that psycho-surgery, although an experimental procedure, could have a therapeutic effect in certain selected cases and that its risks are "not nearly as great as previously supposed." It concluded that there is "at least tentative evidence that some forms of psychotherapy can be of significant therapeutic value in the treatment of certain disorders and the relief of certain symptoms." The Commission recommended further study, and suggested safeguards for obtaining informed consent. It granted permission for operations on involuntarily confined mental patients, prisoners, and children after a court hearing.

Of all the techniques for exercising behavior control, psycho-surgery is the one that should stir up the most apprehension, not because its results are much different from, for example, electro-stimulation, but because it is permanent and irreversible. Once brain cells are destroyed they can never be reconstructed, and whatever changes take place in the individual are there, for better or worse, for the remainder of the person's existence.

Psycho-surgery has been used to eliminate violent behavior in individuals. It has been successful. The concern is that the technique might prove too successful. If it can be used to eliminate unwanted violence, might it not also be used to eliminate aggressive behavior, with "aggressive" being defined as anything violating the gentlemanly norm?

The consideration of informed consent raises its important head again, and because of the type of subject who is a candidate for psycho-surgery the issue is complicated. If the patient is to be subjected to surgery because of abnormal behavior, can he or she give informed consent? Aside from abnormal behavior, the same questions must be asked as would be asked of any medical experimentation when the subject is a prisoner. Because there seems to be little doubt that the operation is of benefit to some patients who cannot be helped by any other means there does not appear to be justification for a total ban on the procedure, as has been suggested by some people. However, because of the type of patient being exposed, and particularly since the effects are drastic, permanent, and irreversible, particular care is demanded in both selecting the patient and protecting his or her rights.

Less violent but equally or more effective for the control of individual reactions and emotions has been direct electrical stimulation by means of micro-electrodes inserted into various parts of the brain. Over the years a great deal of information has become available as to the localization of areas that control various bodily reactions. Considerable evidence, both clinical and in animals, exists to show that when an impulse comes from some part of the body to the brain, the cerebral cortex, the outer layer of the brain, acts only to sort out general impressions, while the real response comes from the inner brain. For example, as early as 1929, Lackey at Harvard University trained rats to travel through a maze. He then progressively removed more and more of the cortex without destroying the learning power of the animal.

Hess, a Swiss neurophysiologist whose work brought him a Nobel Prize in 1949, devised methods for mapping regions deep within the brains of cats by electrostimulation. He inserted, through a hole in the skull, electrodes reaching to the diencephalon, or the "inbetween brain," a region containing among other things the hypothalamus and part of the reticular system. Hess's work established that the diencephalon is the control center for the autonomic nervous system—the system that regulates the working of the heart and other internal organs. Although electrostimulation of the brain surface evokes only isolated muscular movement or twitches, stimulation of the diencephalon actually produced in his cats specific reactions such as abnormal hunger or thirst, panting, sneezing, acting as if they had seen a dog, or even going to sleep.

In 1954 James Olds of McGill University discovered that stimulation of certain regions of a rat's brain seems to invoke an extremely pleasurable sensation. If the rat is allowed to switch on the current by stepping on a pedal, it quickly learns what to do, and spends many hours stimulating itself even in preference to eating.

Dr. José Delgado at Yale University has illustrated the effectiveness of the technique of electrostimulation in a very dramatic fashion. Electrodes were implanted into the head of a bull. The bull was then irritated into a full charge. When signals were transmitted to the electrodes the bull stopped in mid-charge—and peacefully walked away. Without having any direct contact with the animal, Delgado brought it under complete control and kept it that way.

A number of laboratories are now doing the same kind of work on human beings. One experimenter was able to recall to the mind of the subject very specific memories each time a certain portion of the subject's brain was stimulated. In the case of electrical stimulation the process is reversible. But here another problem exists. The patients are totally under the control of another individual, the one who pushes the buttons. The patients' perceptions of the external do not control their actions. They can be made violent or calm without any causative factor except the stimulation controlled by the operator.

There is something about the idea of direct manipulation of the brain that brings to the forefront all of our apprehensions about the results of such experimentation. By whatever mechanism, the brain is the organ that allows us to think. Without a functioning brain we could no longer dream, plan for the future, remember the past, evaluate alternatives, or reason. Our senses—sight, hearing, smelling, feeling, tasting—are mediated through the brain. Our ability to perform physical actions—pick up a stone, kick a football, walk—depends on the proper functioning of the brain.

The brain is an organ of incredible complexity and sensitivity, weighing about three pounds. Most of its mass can be destroyed without in any way affecting the physical or mental characteristics of an individual—providing the correct part is destroyed. On the other hand, the destruction of areas only a millimeter in size can result in convulsions, in loss of sight and hearing, in loss of memory. The areas stimulated by microelectrodes to obtain the reactions mentioned above are located in very specific locations and in small areas. Is it any wonder, then, that tampering with the brain seems to bring up special problems of ethics and to produce a particular sensitivity to direct attack on this organ? I am not arguing for the brain as the sight of a soul or even of a mind. It is sufficient to recognize the brain as a physical organ that allows us to do and to think all of these things that distinguish us from lower animals. One of the reasons direct manipulation of the brain is so serious a procedure is the fact that it is possible to get physical and emotional reactions from an individual while bypassing all sensory inputs through the individual. It is as though the person had been eliminated as part of a selective process. There is no sensory cause stimulating an action. There is only an effect. Therefore, the individual has been eliminated from the process that allows him or her to evaluate something his or her senses have perceived, to make a decision, to accept or reject a conclusion, and then to act.

The composite human—is it here now? Are we approaching the place where, in our search for our roots, our family tree will include the statement, "My mother was a test tube"? Can we look forward to the time when, like a deteriorating machine, a body will be rebuilt as parts wear out?

Let's go shopping for a "Body Model 2001." Choose between the fertilized egg that will develop into the compact five-foot-tall model, or the full-sized six-foot model. Custom-built bodies are available over seven feet tall. Combinations of accessories are numerous. Skin comes in black, white, brown, red, or yellow. Hair is black, white, gray, red, blond or brown. Eyes are black, brown, blue, gray, green and hazel. A good long-term service policy on the body is important. It should be covered for hip-joint replacements, heart and kidney transplants. If some defect does develop in the mechanism, gene replacements should be available. And as the model ages and mental processes slow down there should be a good selection of electrodes available for brain implants.

Miraculous things are taking place in operating rooms right now. Surgery can replace severed limbs or parts of limbs. Prosthetic devices are available that are almost as good as the original. In the sense, then, that an individual may end life with hair, arms, legs, hip joints, heart, or kidneys that were not part of the original body, the composite human is here.

9 . . . *wherein human beings are subjects for experimentation, and the rights of individuals are contrasted to the rights of society*

Scientists are people. Some are competent, some are incompetent, some make mistakes, some are even dishonest. Scientific results are just like other results. Conclusions drawn as a result of today's experiment might not be valid in the light of results of tomorrow's experiment. Quite often when tomorrow's results become available, those responsible for today's results are accused of lack of foresight, carelessness, or even dishonesty.

Nowhere in science is the necessity to modify conclusions based on changing results and new evidence more apparent than in the clinical trial of new drugs, or in the investigation of a new clinical procedure. When preliminary evidence in animals indicates an encouraging trend, the drug is administered to human beings. When the entirely unpredicted and unpredictable side effects show up, the trial is discontinued. Both the animal trials and the human trials were good experiments, and the results were valid. A difference in the results between the animal and human tests does not in any way reflect on the ability or the honesty of those conducting the trials.

What magnifies the problem is the fact that unsuccessful experiments, sometimes harmful experiments, are conducted on human beings. An unhealthy attitude is developing that scientists should be able to guarantee the results of their tests, that absolute safety is an attainable goal. It is not, and can never be.

In the early 1970s the public began questioning many of the practices involved in research on human subjects. In July of 1973 the National Research Act established a commission to identify the basic ethical principles that should govern research involving human subjects; to recommend guidelines and mechanisms for assuring that the principles are observed; to clarify the requirements of informed consent to research in the case of children, prisoners and the institutionalized mentally infirm; and to investigate the use of psycho-surgery and recommend policies for its regulation. The Commission was mandated to investigate the "nature and extent of research involving living fetuses." The Act immediately prohibited all HEW-supported non-therapeutic research on any living fetus before or after induced abortion.

The contradictory aspects of the situation thus created became immediately apparent. Under the decree of the Supreme Court, it was perfectly legal for a consenting mother to have her fetus destroyed for any reason. Under the National Research Act, it was illegal to subject the same fetus to experimentation, regardless of the motives for the experiment or the safety of the experiment. The laws said that under different circumstances the fetus had different rights. In the case of destruction it had no rights of its own, but belonged to the mother. In the case of experimentation, it was considered to have all the rights belonging to any other human being.

As with abortion, one's attitude towards fetal experimentation must be determined by what the fetus is considered to be. If, as some argue, the fetus is simply a part of the mother's body, such as a growth of excess tissue or even an organ, then the only consideration would be the danger or discomfort to the mother. Any experimentation either *in utero* or *ex utero* would be justified as long as it was consistent with proper concern for the mother. If, on the other extreme, the fetus is considered to be human in all aspects, then the same consideration would apply to experimentation whether *in utero* or *ex utero* as would apply to an adult person. The fetus would enjoy all the rights ordinarily assigned to an adult. Only experimentation done for the benefit of the fetus would be justified.

There is a middle view suggested by some that considers the fetus to be neither a glob of tissue nor a human being, but rather like an animal. The respect given to the fetus would be the same as that accorded to any other experimental animal. It should not be subjected to cruelty or needless pain, but, if the experimental results so justify, it may be sacrificed. From the Thomistic point of view the only limit to the experimentation would be that which degrades the experimenter.

What did the National Commission for the Protection of Human Subjects of Bio-Medical and Behavioral Research conclude? General requirements applicable to all research on the human fetus were established:

1. Appropriate prior investigations using animal models and non-pregnant humans must have been completed.
2. The knowledge to be gained must be important and obtainable by no reasonable alternative means.
3. Risks and benefits to both mother and fetus must have been fully evaluated and described.
4. Informed consent must be sought and granted under proper conditions.
5. Subjects must be selected so that risks and benefits will not fall inequitably among economic, racial, ethnic and social classes.

These, of course, are the principles under which any medical research should be conducted.

In their specific recommendations the Commission distinguished between therapeutic research and non-therapeutic research. Therapeutic research is that which directly benefits the subject of the experiment, such as the giving of medication to correct a disease; while non-therapeutic research does not benefit the subject directly, but does result in bio-medical knowledge that can be applied generally. In general any research conducted on a patient not suffering from the disease or condition for which the drug or procedure is being developed is considered non-therapeutic research. The Commission concluded that therapeutic research directed to either the fetus or the pregnant woman could be conducted if the above five general principles are followed. Non-therapeutic research directed toward the fetus *in utero* could be conducted provided that the purpose of research is the development of important bio-medical knowledge that cannot be obtained by alternative means; minimal or no risk to the well-being of the fetus will be caused by the research; and the informed consent of the mother has been obtained and the father has not objected to the research.

Non-therapeutic research directed toward the fetus in anticipation of abortion may be conducted provided such research is carried out within the guidelines for all other non-therapeutic research directed towards the fetus *in utero*. Such research presenting special problems related to the interpretation or application of the guidelines may be conducted provided such research has been approved by a national ethical review body.

As might be anticipated, the consideration of research on the fetus during the abortion procedures or on the nonviable fetus *ex utero* causes the most difficulty. As the Commission points out in its deliberations, such a fetus must be considered a dying subject. By definition, therefore, the research is non-therapeutic, because the subject of research will receive no benefit from it. Regarding this research the Commission recommended that it be allowed to proceed if: the purpose of the research is the development of important bio-medical knowledge that cannot be obtained by alternate means; investigation on pertinent animal models and non-pregnant humans (when appropriate) has preceded such research; the research has been approved by existing review procedures; the informed consent of the mother has been obtained, and the father has not objected; the fetus is less than 20 weeks gestational age; no procedural changes are introduced into the abortion procedure in the interest of research alone; and no intrusion into the fetus is made which alters the duration of life. Such research presenting special problems related to the interpretation of these guidelines may be conducted provided such research has been approved by a national ethical review body.

Non-therapeutic research directed toward the possibly viable infant may be conducted using the same principles applied to any other human research.

In accepting the report of the Commission, the Department of Health, Education and Welfare departed from the recommendations in two major areas. HEW introduced a provision that would allow the rules to be waived in certain unspecified instances. The Commission recommended a national review body that would aid in interpretation, but did not suggest the possibility of waiving regulations.

In another major area, HEW did not accept the specific recommendation that, in non-viable fetuses, prolongation of life not be allowed. Instead, the Department approved this research saying that such research contributed to the ability of the physician to bring viability to increasingly small fetuses.

This brings up a very interesting point. Suppose the fetus is declared non-viable, then placed on a respirator and kept alive for the sake of experimentation. Is it morally or even legally permitted to take the fetus off the respirator—to pull the plug—when the experiment has been completed?

The whole relationship of a Commission to the Department of HEW is illustrative of the difficulty in trying to judge science as moral or immoral and of making judgments based only on scientific desirability rather than on ethical and moral grounds. The eleven men and women making up the Commission represented a wide spectrum of different backgrounds and philosophical outlooks. They included the chairman of a department of obstetrics and gynecology, a professor of behavioral biology, and vice-chancellor for health services of a major university, the president of the National Council of Negro Women, a Jesuit professor of bio-ethics, two professors of law, a professor of Christian Ethics, a chairman of a department of internal medicine, a provost of a major university who is also professor of physiological psychology, and a practicing attorney. Thus, the recommendation either on scientific or on moral and ethical grounds or both should not be taken lightly. Yet HEW chose to disregard a recommendation based on ethics in favor of its own decision based only on scientific desirability.

The hardest points to rationalize in the recommendation of the Commission were those relating to experimental work on a fetus, either *in utero* or *ex utero* when a decision to abort had been made. It is pretty hard to justify abortion on the one hand by saying that the fetus is not human and has no rights, and then, on the other, to limit the kinds of research that can be done on that nonhuman tissue that has no rights.

Several related questions have been confused in the discussion on fetal research. The question "is the fetus a person" is not the same as "is the fetus viable" nor is either the same as "is the fetus human." The decision as to whether or not the fetus is a person defines what legal rights the fetus has. If it is a person, then it has all the rights provided by law for the protection of an adult. The Supreme Court has decided the fetus is not a person. Therefore, it has no rights under the law.

The question as to whether or not a fetus is viable is scientific and medical. The continuing difficulty I find with making decisions based on the viability of the fetus is that its potential for viability will vary at different times depending on the technical skills available for its preservation. Viable is usually understood to mean that the fetus is capable of growing and developing outside the uterus. Thus, as technology and medicine advance, younger and younger fetuses can be brought to this stage, and so specifying a time limit for viability is an ever-changing process. In the broadest sense of the term "living," the zygote is alive from the moment of fertilization, since it then becomes a tissue, capable of dividing and reproducing. It is not living in the sense that living is the opposite of dying or dead, since human death has been described by the absence of certain physiological responses that the zygote never had, and that the fetus will not develop until after weeks of gestation.

The requirements for fetal research specified by the National Commission are the same factors that must be taken into account when performing experiments on any human being. Every experiment is potentially dangerous; therefore, there should be an important reason for doing it. It must benefit the patient or add to medical knowledge that will benefit other patients. Risks must be weighed against benefits.

To illustrate the difficulties of weighing risk against benefit, and then deciding on ethical factors, let me give you an actual example of an experiment. About twenty-five years ago—in the early 1950s—doctors at a Boston hospital treating diabetic women who usually had difficult pregnancies thought they had evidence that showed that Stilbestrol (DES), a synthetic chemical possessing the properties of the female sex hormone estrogen, could prevent spontaneous abortion. The product had been in use in human beings for some years before this time for other medical uses, and no adverse side effects, under proper conditions of use, had been observed. The problem of spontaneous abortion is a serious one. Many women, desperately wanting children, are able to conceive, but miscarry before the fetus reaches the self-survival stage. When a woman has had three or four or even as many as nine or ten such miscarriages she is classed as a "habitual aborter." If in further trial Stilbestrol confirmed the early clinical impression, it should then be possible to treat women in early stages of pregnancy, allowing them to deliver a baby at normal term. It was also believed that the chemical could be used to diminish the number of premature babies for the same reason.

Although a number of normal children were born to women participating in this program, a statistical analysis of the results after many years of use showed that the effect of Stilbestrol was not enough to justify making a claim for the prevention of abortion, and its use was discontinued. No adverse side effects were caused in the women by the Stilbestrol, although the administration of the drug had been approved medical practice for over twenty years.

Then, in the early 1970s, an analysis of data obtained on women in their early twenties showed a startling association of cause and effect. Data showed that daughters born to mothers who had been treated with Stilbestrol for threatened abortion had a higher incidence of a rare cancer of the vagina than did the general population.

So far only about 350 such cases have been detected, about one-third of which were found in women with no known exposure to DES. If the disease is diagnosed early it usually can be treated successfully.

However, as many as one-third of the daughters may have a condition of the vagina known as adenosis. It was feared that this was a pre-cancerous condition. It appears now that this condition disappears spontaneously by the time a DES daughter reaches her late twenties. The one remaining question is whether or not a cancerous condition will develop when the daughter reaches menopause.

The hard questions now arise. It was assumed that the drug would prevent abortion. There was no way of knowing without experimentation that it would not. If it had been known definitely that the treatment would subsequently produce the type of vaginal cancer in 0.01 percent to 0.1 percent of the females born of mothers who had taken Stilbestrol should it have been used? If you knew that your daughter would be one of the one in a thousand or one in ten thousand who would get cancer, and of those she would be one of a small number of incurables, would you consider a life of twenty-five years without cancer better than no life at all, or would it have been better never to have been born? If you, as a habitual aborter or the husband of a habitual aborter, had been asked for informed consent under these conditions would you have given it?

In referring to these results, an editorial in a major newspaper commented that some of the hospitals involved were accepting the moral responsibility for "mistakes" made in the past and following up on the daughters of all treated mothers. Was the attempt to prevent habitual abortion a "mistake" because it wasn't successful? Obviously if it had been known that the drug was not effective, the experiment would not have been done. But, knowing everything else, even in retrospect the decisions on physical risk and benefit are not easy. The ethical and moral decisions are even harder.

The first to report on a possible link between women who took DES and the appearance of later vaginal or cervical cancer in their daughters was Dr. Arthur L. Herbst, then at the Massachusetts General Hospital, who observed seven or eight cases of very rare carcinomas in females between the ages of fourteen and twenty-two. This was a far higher incidence than would have been expected in this age group. Following wide publicity of this report, clinics in various sections of the country began to carry out studies on daughters born of mothers who had taken DES to prevent miscarriage. By 1977 Dr. Herbst had recorded 154 DES-associated cases of cancer in women born between 1948 and 1956. By 1977, the situation had landed in the courts. Both mothers and daughters were suing pharmaceutical manufacturers and hospitals. One of the lawyers for the plaintiff called the case "the biggest scandal since thalidomide."

However, in May of 1977, Dr. Herbst, now at the University of Chicago, in association with researchers at the Harvard School of Public Health, Dartmouth Medical School, and the Massachusetts General Hospital, published a follow up article. He estimated the risk of cancer at a level lower than what earlier publicity had led the public to believe. He said that the odds of a daughter developing cancer ranged from a low of about 1.4 in 10,000 to a high of 1.4 in 1,000. Many of these cases would be curable.

In March of 1978 the Secretary of HEW created a task force to review the entire problem. The report was submitted to the Secretary in September 1978 and, on the basis of its recommendation, an advisory letter was sent to every M.D. and osteopath in the country. The task force concluded that between 4 and 6 million people were exposed to significant doses of DES. This includes mothers for whom DES was prescribed, as well as for their sons and daughters. The incidence of clear cell carcinoma in the daughters is "well established" but small, probably between 1.4 per thousand and 1.4 per ten thousand in the exposed population through age 24.[1]

The legal aspects of the case present many difficulties. The mothers are suing because they were not told of the risk involved. However, there were no known risks, and the risks could not have become apparent for fourteen to twenty-two years. There is no suggestion that the mothers themselves suffered any serious side effects.

The final irony is that it is probable that at least some of the young women would not be alive to sue if it were not for DES. DES had been standard medical practice for over twenty years in preventing spontaneous abortion. With the discovery of the cancer association, approval for DES use for this purpose was withdrawn. However, it still cannot be said that DES is *useless*. What must be said is that there is no statistical evidence to prove that it is *useful*.

There is a tremendous difference between these statements. The FDA requires statistical proof that a drug is efficacious before it can be approved for use. The belief of a physician or hundreds of physicians is not enough to justify approval. In dealing with DES the difficulty lies in not having a statistical base on which to form a judgment. If a woman has had several spontaneous abortions, there is a possibility that she will have another if she becomes pregnant. However, it is not certain. If DES is administered, and if she gives birth to a normal child, there is no way of determining whether or not DES was of help. In an article in the journal *Cancer* in 1973, Dr. Heinonen estimated that between 100,000 and 160,000 live born female infants were exposed to DES between 1960 and 1969. The incidence of use during the 50s was much higher. Therefore, it can be said that hundreds of thousands of children were born to women who were so-called habitual aborters, or who had threatened abortion during pregnancy, and who took DES to prevent abortion. Even though statistics cannot be used to support the belief, testimonial evidence from many physicians who trusted DES for over two decades would lead to the conclusion that at least some abortions were prevented by DES.

Another example of the ethical and practical difficulties surrounding research on humans is illustrated in a study reported in late 1975 on children suffering from phenylketonuria (PKU). The phenylalanine build-up in the body results in mental retardation. If this defect is detected in the first few weeks of life, the child can be placed on a phenylalanine-free diet, and will develop normally. This fact has been known for some years. However, it is not known how long the child must remain on the diet. It is extremely difficult for the child to adhere to a diet as he or she grows older. It is also expensive.

Researchers at Johns Hopkins have conducted a small scale experiment involving ten PKU children. At the age of four, five of the children were transferred from the phenylalanine-free diet to normal food, while the other five were kept on the diet. The children were followed carefully, with both physical and psychological tests being administered for two years. At the end of that time there was no statistically significant difference in the IQ of the two groups.

The ethical problem is obvious. Taking a child off a diet might cause irreparable mental damage. The benefit would be that the child could lead a normal life and the parents would be released from a significant financial responsibility of seeing that the child did not violate the special diet by eating harmful foods.

A different ethical dilemma arose during the course of a clinical trial of a drug being tested as a possible preventive of the sudden death that sometimes occurs within the months after a patient has suffered a heart attack. It had been observed that a drug that had been on the market since 1959 as a treatment for gout prevented red blood cells from forming clots. As a result of this observation it was decided in late 1977 to run a trial on patients who had recently had heart attacks. As is customary, one group of patients received one drug, a similar group received a placebo, and neither the doctors nor the patients knew which treatment an individual patient was receiving. The trial was planned as a year study. However, after about eight months of treatment, scientists monitoring the tests observed results that were so positive it was apparent that the drug really was effective in preventing further attacks. The question arose as to whether or not it was ethical to continue the treatment of the group on the ineffective placebo, knowing that an effective drug was available, even though the patients in the trial had signed a consent form knowing they might be part of the placebo group. The decision was made to notify the patients and their doctors of the preliminary results so that the patients could decide whether or not to withdraw from the study. It is difficult to believe that a patient would agree to continue on a placebo knowing that a lifesaving drug is available.

It has been estimated that if the preliminary results could have been confirmed by the continued trial, the use of the drug would prevent 200 to 300 deaths a week. Of course, stopping the trial does not mean that the drug cannot be used in the future. It does mean that the research results necessary to confirm its usefulness cannot be obtained.

But the correct ethical decision was made. The importance of obtaining the scientific data did not justify the risk to the individuals involved. Only the individual could decide whether or not to take that risk.

Periodically controversy occurs over the source of patients or subjects used for human experimentation. It has become common over the years to use prison populations in clinical experimentation. Most of these trials involve what would be called non-therapeutic research, research that would not benefit the subject directly. For example, if we needed to know how much of an antibiotic was absorbed into the human bloodstream at a particular dose level, or if it were desired to know how a drug is metabolized in the body, it would not be necessary to give the drug to a person suffering from a disease. Most such trials are run on normal people and, on the basis of the results obtained, the appropriate drug level for treating a disease is determined.

The opposition to using prisoners comes not from them but from those concerned with civil liberties. Their thesis is that a prisoner cannot give informed consent; that, because of the advantages accruing to the prisoner as a result of being a part of a clinical research program, he or she is forced to consent. Certainly, advantages accrue to the prisoners. In states or other localities where authorities stopped all prison research the prisoners themselves protested the decision. They pointed out that they were unharmed, that the money they received helped them pay off fines and court costs and, in many situations, allowed the prisoner at the end of the sentence to enter the free world with more than the state's discharge grant. As one prisoner expressed it, the grant alone would be "enough to buy a gun and a few bullets," but the addition of the research pay would be enough to give a fresh, legitimate start.

Suggestions have been made that scientific review groups such as exist in research hospitals be established to protect the prisoner, and also that the subjects be compensated for all lasting injury or loss of earnings suffered as a result of drug experiments. Both of these conditions are quite commonly, if not universally, met today.

Another suggestion has been that prisoners be paid the same amount as would be required to attract a non-prisoner into the same research project. This is a ridiculous suggestion on the face of it. If payment is a coercion to a prisoner, how can a higher payment be more protective? Clinical trials are not run in prisons because the trials are less expensive there. The payment of volunteers is a very small fraction of the cost of a trial. Prisoners are selected as subjects because they are part of a group whose habits, diet, and exercise are known and can be observed. In addition, the control group, the group not getting the drug but who are used for comparison, have the same habits. There is no chance, for example, that someone in the experiment will end up on a three-day drunk, or eat something that would destroy the results of the test, or might unknowingly even take another drug.

I do not believe that rewards for participation destroy the dignity of a volunteer—prisoner or non-prisoner. Despite the horror tales, and some of them are real, of experiments conducted some years ago on both prisoners and non-prisoners, prison populations are given the same treatment, including safeguards, information, and warnings, as are subjects in the finest research hospitals.

In 1974, primarily in response to publicity about the experiments on syphilitic inmates in the Alabama State Prison at Tuskegee, and the disagreements over fetal research, Congress established the National Commission for the Protection of Human Subjects of Bio-Medical and Behavioral Research. As part of its charge the Commission was directed to propose guidelines for the federal regulation of research on certain special populations of research subjects: human fetuses, prisoners, institutionalized psychiatric patients, children, and other captive groups. In October of 1976 the Commission issued a report that included its recommendations for the use of prisoners as experimental subjects. The recommendations are:

(1) Studies of the possible causes, effects and processes of incarceration and studies of prisons as institutional structures or of prisoners as incarcerated persons may be conducted or supported, provided that (a) they present minimal or no risk and no more than mere inconveniences to the subjects, and (b) the requirements under recommendation (4) are fulfilled.

(2) Research on practices, both innovative and accepted, which have the intent and reasonable probability of improving the health or well-being of the individual prisoner may be conducted or supported, provided the requirements under recommendation (4) are fulfilled.

(3) Except as provided in recommendations (1) and (2), research involving prisoners should not be conducted or supported, and reports of such research should not be accepted

by the Secretary, DHEW, in fulfillment of regulatory requirements, unless the requirements under recommendation (4) are fulfilled and the head of the responsible federal department or agency has certified, after consultation with a national ethical review body, that the following three requirements are satisfied:

(a) The type of research fulfills an important social and scientific need, and the reasons for involving prisoners in the type of research are compelling;

(b) The involvement of prisoners in the type of research satisfies conditions of equity; and

(c) A high degree of voluntariness on the part of the prospective participants and of openness on the part of the institution(s) to be involved would characterize the conduct of the research; minimum requirements for such voluntariness and openness include adequate living conditions, provisions for effective redress of grievances, separation of research participation from parole considerations, and public scrutiny.

(4) (a) The head of the responsible federal department or agency should determine that the competence of the investigators and the adequacy of the research facilities involved are sufficient. . .

(b) All research involving prisoners should be reviewed by at least one human subjects review committee or institutional review board comprised of men and women of diverse racial and cultural backgrounds that includes among its members prisoners or prisoner advocates and such other persons as community representatives, clergy, behavioral scientists and medical personnel not associated with the . . . research or institution. . . .

(5) In the absence of certification that the requirements under recommendation (3) are satisfied, research projects covered by that recommendation that are subject to regulation by the Secretary, DHEW, and are currently in progress should be permitted to continue not longer than one year from the date of publication of these recommendations in the Federal Register (Jan. 14, 1977) or until completed, whichever is earlier.[2]

The conditions requiring "compelling" reasons for the use of prisoners, equity, and high degree of voluntariness on the part of the prisoners and "openness" in the prisons virtually assure the elimination of the further use of these subjects, since these are the very conditions that are subject to the most vigorous debate on the meaning of the terms.

The Commission did succeed in collecting data that indicated the extent of such experimentation. The Pharmaceutical Manufacturers Association, whose members develop most of the prescription drugs approved in this country, reported that three-quarters of its members, operating in eight states and six county or municipal prisons, used 3600 prisoners as subjects in 1975. Between 1970 and 1975 five of the six agencies in the Public Health Service conducted 124 biomedical studies and 19 behavioral studies in the U.S. prisons. In addition, the Department of Defense and the Atomic Energy Commission also sponsored such research.

The Commission also found that research on prisoners is limited to the United States. This is not surprising, since no other country, with the exception of Italy, requires the testing of drugs on normal subjects before conducting clinical trials or therapeutic trials. Italy requires a one to three month study on normal, healthy volunteers before a drug can be tested for the condition for which it was being developed.

The overriding factor resulting in the Commission's recommendation seemed to be the belief that prisoners could not give informed consent. With this as a base, it is obvious that no other consideration could change the balance of argument in favor of experimentation. The utility of the program was agreed to. Even the consideration of risk did not seem to be too much of a negative factor. The Pharmaceutical Manufacturers Association submitted evidence that indicated that not a single prisoner had died or been permanently injured as the result of a drug firm-sponsored test, and that serious toxicity rarely occurs in such tests. The Commission itself agreed that "the risks of research, as compared with other kinds of occupations, may be rather small" even within prisons. In a survey published in the *New England Journal of Medicine* in September 1976, the conclusion was reached that "The risks of participation in non-therapeutic research may be no greater than those of everyday life, and in therapeutic research, no greater than those of treatment in other settings."

Of the 93,000 subjects in non-therapeutic research covered in this study, 0.8 percent were reported injured, no one died, one subject was permanently disabled (although some doubt that the subject's stroke was caused by the research), thirty-seven were temporarily disabled (reactions to drugs, corneal abrasions, electrical burns, and assault by another participant), and 673 suffered trivial injuries (discomfort, scars, colds, and mild allergic reactions). Of the 39,000 therapeutic research subjects, 10.8 percent were reported injured; forty-three deaths, thirteen permanent disabilities, 937 temporary disabilities, and 3,253 trivial injuries were reported.

The Commission visited one prison and talked with the prisoners who had participated in experiments and those who had not. It was reported that many were indignant at the idea of research being discontinued. A study performed by the Survey Research Center at the request of the Commission found that 87 percent would be "very willing" to participate in another research project. If, however, one believes that prison conditions prevent a subject from giving informed consent, and if, as it should, the Commission had as its goal "the principle of respect for persons, which require that autonomy of persons be promoted and protected," then, inevitably, the conclusion must be that prisoners are not suitable subjects for research. The other issue that resulted in a major recommendation was that of "equity." The argument was used that the prisoners were taking a disproportionate amount of risk compared to the benefits they received.

I find both of these points singularly unconvincing. The stimulus for anyone to volunteer to be a part of a non-therapeutic medical experiment will be basically economic. Granted, there will always be a small number who will volunteer out of a desire to advance medical knowledge and help others, but this will be a small, probably inconsequential, percentage of those necessary to assure the continuation of research. People will volunteer because they need money. Of those volunteering by far the highest number will come from the lowest economic level of our society. It is not easy to believe there will be an equity distribution among members of the various economic classes in our country. They will not be prisoners of the state, but they will be prisoners of their economic conditions. Just as incarcerated prisoners will volunteer, not because they want to, but because they need additional resources, so too will nonprisoners. In fact, it could be argued that there is less coercion used on prisoners than on non-prisoners.

The prisoners are still assured of meals and shelter while in prison whether they volunteer or not. Non-prisoners sometimes volunteer because the income they receive is necessary for their existence.

I don't accept the thesis that payment to prisoners for participating in experimental programs is "coercion." If they were not paid, the prisoners would not volunteer. If a financial reward for doing something an individual does not want to do is coercion, then probably a large percentage of our population is being coerced. It is obvious that people doing hard physical labor, those doing menial jobs, those working under uncomfortable conditions do so because they are getting paid. It does not seem to me that it can be argued that these individuals are not exercising informed consent when they accept a job they do not want since they accept it because they must if they want to be paid.

Dr. Benjamin Freedman, Professor of Ethics and Law at Tel Aviv University, explains in a clear way the distinction between rewards for performing a service and coercion to perform a service. He points out that there are certain basic freedoms and rights that entitle each individual to certain things. If in return for service we are given a reward above those rights and freedoms, it is a true reward. If on the other hand, the reward only brings the individual up to the level of the rights and freedoms to which he or she is entitled anyway then the reward takes the form of a threat—"If you don't do this, you will not eat"—that is true coercion.

In the prison context, for example, it would be no real reward to allow prisoners to learn a trade, or to get adequate health care, since they are entitled to these things. It would be a reward to increase their daily pay above the minimal level, since they are not entitled to that.

The argument against requiring an equal distribution among the various classes of society in order to get proper representation in such a volunteer group follows from this. It is unrealistic to suppose that people in upper income brackets would become part of such experimentation. Therefore, it is inevitable that those who will bear the major part of such experimentation are those who need money and, possibly, cannot obtain it any other way.

The same general argument can be used when considering therapeutic trials. There is one difference, and that is the possibility that the subject of the experiment would gain direct benefit from it. For example, individuals from all classes who were suffering from a particular disease that could be cured as the result of the trial would volunteer.

I can see nothing wrong with this system. In our present atmosphere all possible precautions are taken to protect the patient. He or she is guaranteed against future loss resulting from the experiment. Under such circumstances the whole procedure could be considered in the nature of a job. When the government creates programs to give employment to 100,000 unemployed it is considered a humane gesture. Many of the jobs do nothing to increase the dignity of the individuals, other than allowing them to support themselves. If the dignity of the individual is to be a major concern, and it must be, I consider an activity that advances medical knowledge with the possibility of helping thousands of suffering people to be a far better choice than making license plates.

A separate problem occurs when it is necessary to experiment with children. To test a drug that is to be used in the cure or prevention of a childhood disease, the drug must, at some point, be administered to children. In fact, the FDA requires that drugs be tested over the entire age range for which they are to be recommended and used. The added ethical difficulty is that children themselves cannot give informed consent nor can they understand or appreciate the risk to which they might be exposed.

A classical example of a case that was susceptible to widely differing ethical interpretations occurred at the Willowbrook State School in Staten Island in the early 1970s. Willowbrook is an institution for the mentally retarded. As was common in this institution, serum hepatitis affected most of the children. A patient usually developed at least a mild case of hepatitis within the first year of admission. Because of the prevalence of the disease, and because the population of the institution was under direct observation, a research project was started to develop an anti-hepatitis vaccine. All the usual precautions were taken. The parents of the children were completely informed, and gave their consent.

The program involved the development of a vaccine based on the attenuation or inactivation of the disease-causing organism. The virus-like particle is inactivated to the point where it can no longer cause the disease, yet it retains properties that stimulate the body to develop antibodies to the pathogenic organism. Thus the body builds up immunity to the disease. Blood serum from those patients who were known to be affected was treated. Then ten newly-admitted patients were given one injection of the vaccine, and another four were given two doses. After waiting for several weeks to give the vaccine a chance to stimulate immunity, the children were each given an injection of serum containing the hepatitis-producing agent. Of the ten children who had received one dose of the vaccine, five were protected, but the other five showed symptoms of the disease. Of those four receiving the double injection of vaccine, all were protected.

So it appeared that here was a vaccine that could prevent the hepatitis that affects 150,000 to 200,000 people from blood transfusions each year, and save the 3,000 people who die from the same cause. However, there was an immediate outcry both here and abroad regarding the ethics of the experiment. Since the fourteen children did not have hepatitis when they became part of the experiment, critics charged that the experiment was therefore not therapeutic, that is, the children themselves could derive no benefit from it. Proponents answered that since the children would eventually get the disease anyway, anything that could be done to develop a vaccine would certainly be of benefit to the individuals involved.

This was not the first such experiment done at Willowbrook. An almost identical procedure was used to develop measles vaccine. Measles, like hepatitis, was also prevalent at Willowbrook. In 1960 it accounted for sixty deaths. For some years following the advent of the vaccine in 1963 there has not been a single case of measles in the institution. The results of this work extend far beyond the Willowbrook State School. Countless thousands of children have been spared the effects of measles as a result of this work.

It was over one hundred and eighty years ago that Dr. Edward Jenner innoculated James Phipps, then only eight years old, with pus taken from a patient's cowpox sore. After seven weeks the child was innoculated with smallpox, a disease for which there was no known cure at that time, and from which about half of its victims died. James Phipps did not contract smallpox.

That single experiment was the basis for one of the most dramatic advances of medicine. Because the experiment was a success, and because it resulted in almost incalculable good, the ethics of the experiment are not often questioned. Yet James Phipps could not give informed consent nor, I suppose, did anyone give informed consent for him. He was not suffering from smallpox, and so could not benefit from the experiment. In the light of today's standards there was not even scientific justification for performing the experiment.

But the ethics of an experiment do not depend on whether the experiment is a success or a failure. An experiment is ethical or unethical from its inception. The dilemma is apparent. It is insolvable, and can only be approached by using every possible precaution when it is determined necessary to perform experiments with children. And it is necessary if the afflictions of children are to be relieved or eliminated. Yet there is always the nagging doubt that even the consent of loving parents is given because these are only children, and therefore more expendable than adults. In particular, the use of mentally retarded children must be undertaken with caution and compassion. Possibly it is unfair to ask if parents might give the same consent for the same experiment to be performed on their normal child.

On August 19, 1947, sitting as an international tribunal at Nuremberg, three American jurists pronounced a judgment on what should be required for human experimentation. Nuremberg rule number one states: "The voluntary consent of the human subject is absolutely essential."[3] The same injunction was issued in other codes. The World Medical Association in its "Principles for Those in Research and Experimentation," issued in 1954, said: "It should be required that each person who submits to experimentation be informed of the nature of, the reason for, and the risk of the proposed experiment. If the patient is irresponsible, consent should be obtained from the individual who is legally responsible for the individual. In both instances, consent should be obtained in writing." The Declaration of Helsinki, issued in 1964 by the World Medical Association, says: "If at all possible, consistent with patient psychology, the doctor should obtain the patient's freely given consent after the patient has been given a full explanation. In cases of legal incapacity consent should also be procured from the legal guardian; in cases of physical incapacity the permission of the legal guardian replaces that of the patient." The AMA Ethical Guidelines for Clinical Research states, in three different places, that: "The voluntary consent of the person on whom the experiment is to be performed should be obtained." The Department of Health, Education and Welfare, in the *Federal Register* of May 30, 1974, published its final regulations titled "Protection of Human Subjects Relating to Human Experimentation.'' They define human consent as:

> the knowing consent of an individual or his legally authorized representative so situated as to be able to exercise free power of choice without undue inducement or any element of force, fraud, deceit, duress, or other forms of constraint or coercion. The basic elements of information necessary to such consent include:
>
> 1. A fair explanation of the procedures to be followed and their purposes, including identification of any procedures which are experimental;
> 2. A description of any attendant discomfort and risks reasonably to be expected;
> 3. A description of any benefits reasonably to be expected;
> 4. A disclosure of any appropriate alternative procedures that might be advantageous to the subject;
> 5. An offer to answer any inquiries concerning this procedure; and
> 6. An instruction that the person is free to withdraw his consent and to discontinue participation in the project or activity at any time without prejudice to the subject.

With such unanimous agreement that informed consent is essential why is there so much furor around this issue? In most of the cases that have been brought to court, the problem usually revolves around the fact that, although consent was given by the patient, the information relating to the risk was not sufficient to allow the subject to make an informed decision on the risk versus the benefit. The complaints are probably justified in some cases. However in many, the information that would have been required was not available, and could not have been obtained without the experiment. The accusations, ironically, are based on the belief that the bad effects of the experiment should have been predicted before the trial was run. One such instance, mentioned previously, was the administration of DES for the prevention of miscarriage. It was not until twenty-five years after the trial that the results on the daughters of the patients became available. Another similar case involved the use of anticoagulants, medicines to prevent the blood from clotting, in cerebral-vascular diseases. Doctors reasoned that if the blood could be made thinner fewer strokes would result. It was considered good medical practice to administer these drugs to patients at risk. However, when a huge, long-term control study was run, the results indicated that not only was the treatment ineffective, but the anticoagulant actually seemed to increase the risk of a cerebral-vascular accident. If we put ourselves in the position of the physician, it would have been unethical to have withheld such treatment when the body of medical opinion said it was of benefit. However, it would be equally unethical to prescribe the drug now that the study has indicated side effects.

A second reason for a reluctance to obtain informed consent is that, particularly in a blind study, a knowledge of what might happen could influence the reaction of the patient to a drug. In a blind study the patient does not know whether he or she is getting an active drug or an inactive material (a placebo). In a double blind study neither the physician nor the patient knows whether an active or inactive material is being administered. The so-called "placebo effect" is well-known and accepted. If a patient is told that a particular effect might be experienced, in a certain number of cases he or she will experience the effect as a result of a psychosomatic reaction—thus, the reluctance to enumerate all the effects that might be experienced. However, this effect can be neutralized in a control study. If the effect is real, it should occur more often in the group getting the active material than in those getting the placebos, even though some psychosomatic reactions will occur in each group.

The third reason for not obtaining informed consent is that a recital of the possible risks might cause a patient to refuse treatment or to refuse to be part of an experiment. I believe this factor is greatly overrated.

In 1971 Dr. Ralph Alfidi of the Cleveland Clinic set up a study to determine the reaction of patients to a complete disclosure of the risks of a diagnostic procedure. The procedure selected was the performance of an angiogram, a procedure that involves insertion of a catheter into one or several blood vessels, the injection of an opaque dye through the catheter, and finally, the use of X-ray to determine any abnormalities in the vessels. A form was prepared outlining the procedure and then listing the possible, even though rare, side effects in rather harsh language. The form stated, for example, ". . . although the risk is very small, clotting the blood supply to an organ can result in the loss of that organ, and, remotely, in the loss of life. The latter is true of other complications, but is just as rare," and, "during the procedure it is possible that dye (contrast medium) might result in an adverse reaction causing hives, shortness of breath, extremely low blood pressure, and, rarely, temporary or permanent paralysis." In only two cases of 132 did such complete disclosure cause the patient to reject the procedure. When a second but similarly harsh form was used the results were comparable—only two patients out of 100 refused.

As in other studies the results of this investigation show that the patient wants to know what is happening or is going to happen. Most people now recognize how complex medicine is, and take almost for granted that there will be some risk. Given the right information the patient should be trusted to make the proper decision.

One of the minor controversies arising as a result of the consideration of the rights of the patient concerns the ethics involved in prescribing a placebo. There can be no doubt about its effects. The placebo has been a part of medicine for centuries, particularly in relief of pain. About 30 to 40 percent of cancer patients report some relief when given a placebo and told that it is a powerful analgesic. Published studies show that subjects can be made to become euphoric or nauseated at the suggestion that the medication they are taking has these effects. Even color, shape and size of the pill cause different effects. A study performed in England concluded that red pills were more effective than blue or green ones, and that yellow had the least effect in relieving the pain of arthritis. Green is the preferred color for treating anxiety, and yellow is best for depression. There is now a "long-acting placebo" consisting of multi-colored small capsules encased in a standard sized capsule to mimic the appearance of a popular long-acting drug.

The objection to placebos is based on the ethical validity of practicing deception. I believe deception is too harsh a word here. The stimulation of the mental attitude of a patient is part of the practice of medicine. Placebos are never used, for example, in trying to cure infectious disease. They are used only when it is desirable to relieve the stress of an individual, and so eliminate symptoms such as headache or other pain. The word placebo is derived from the Latin meaning "I please." If the patient is pleased, stress will be relieved and symptoms will disappear. Placebos can be justified on a risk–benefit analysis. Certainly it is always less dangerous to administer a material that causes no organic, or physiological, reaction than it is to give a material that does produce some measurable effect.

The second argument against the use of placebos is that the discovery by the patient that he or she was being given an inactive drug would affect the doctor–patient relationship. In order for the drug to be effective the patient must be told by the attending physician that it will cause a certain effect. The physician must lie to the patient. The British Medical Journal, *Lancet,* has taken a strong position on the use of placebos, even while admitting they could help as many as 30 percent of patients. They ask: "If 30 percent of all National Health Service prescriptions were pharmacologically inactive placebos rather than active ones, would our patients any longer have faith in our medicine when rumors spread that the medical profession was deceiving them into a cure?"

It is an interesting dilemma. If the physician uses the placebo, the patient might discover that the drug was inactive, and the treatment would then become ineffective. The alternative would be no treatment or, if a physiologically active drug were available, use of a material that might cause other effects. In this situation I can find no ethical difficulty. The effect of the placebo depends on the physician–patient relationship. The physician, therefore, should judge the extent to which he can use suggestion to help the patient. It is always better to use a non-physiologically active material, assuming the results to be the same as those obtained from an active drug if one is available. The "deception," therefore, is in this case justified, not as a pernicious practice, but rather as part of the art, rather than the science, of medicine.

Some agreement on humanness must be reached if we are to decide what is morally and ethically right in the way of human experimentation. On one hand there is the extreme position that the scientist has the right to do anything technique and technology make possible. Many scientists believe this. On the other extreme is the position that a scientist has no moral right to intervene in natural processes. Practically no one believes this. What I believe should be proposed is a position that says that a scientist has the moral right to intervene in nature to enhance those qualities that make a person more human, and to eliminate those qualities that make for less humanness.

It is possible to list those things that make humans unique, that distinguish us from other animals: speech, ability to reason, etc. An attempt could even be made to list those things we value most, and what objectives we wish to accomplish. Aristotle would list *eudemo-*

nia or happiness, the hedonist would list pleasure. Herman Muller says that people most generally value, among other things, "a genuine warmth of fellow feeling and a cooperative disposition, a depth and breadth of intellectual capacity, moral courage and integrity, an appreciation of nature and of art, and an atmosphere of expression and of communication."[4] The founders of America would include "life, liberty, and the pursuit of happiness." And I would add my own, with Zeus, "justice and a reverence for others." The ancient Hebrews depended on revelations from God to Moses. Christian thinkers, in seeking the *telos,* the final goal of humanity, departed from the Bible in including the "natural law," a law reflected in the natural tendency of people to do what is good. If, by a rational approach, conclusions could be drawn from a naturally occurring world order in the universe on what we should do, then the essentials of human beings could be determined, and finally, what we ought to be normatively.

If medicine is to progress, if new preventions and cures for diseases are to be discovered and developed, human experimentation must take place. The only way to uncover the effects of a drug on human beings is to administer it to human beings. All the animal tests in the world cannot predict with certainty the benefits or side effects of a drug.

The most extensive drug testing imaginable on human beings still does not predict what untoward reaction might be encountered in other human beings. Therefore, every administration of a drug, even an established drug, is in the nature of an experiment. Every individual has the potential to react in a way different from that seen in thousands of persons treated previously, because each individual is unique. It follows, then, that there is risk associated with every procedure.

If truly informed consent really exists, then all ethical problems disappear. I include as a requirement for informed consent lack of coercion and total disclosure. Under these circumstances the subject of the experiment is free to act regardless of the risk to which he or she might be exposed. If reward is interpreted as coercion, then informed consent happens rarely. If it is necessary to be able to predict everything that might happen as the result of an experiment, there can be no informed consent. If there is no informed consent, there can be no progress in medicine.

10. . . *wherein it is shown that we are as gods, and now need to learn to act as gods*

For the first time since human beings became human we can think of designing our future. It is a heady thought, not just to be able to design our environment, but to design ourselves.

I have shown what can be done. Should it be done?

It is difficult to quarrel with a belief that says that what is of benefit to humankind should be done, and what brings harm to humankind should be avoided. However, it is too simplistic to agree totally with these statements. What makes the statements simplistic is the belief that there is agreement on what a "benefit" is and, even worse, that there is agreement on what "humankind" is—or should be.

I have outlined the scientific and ethical disagreements brought about by our changing science and technology. We have seen a dramatic retreat from the Hegelian principle that "what is useful is right." In some areas what would have been looked upon as a real contribution to a better life, and accepted with alacrity a quarter of a century ago, is now looked upon as a danger and, if accepted at all, is done so only with some suspicion and after thorough investigation. If I had to select one reason for this change in attitude I would say that it was caused by a new appreciation for the rights of the individual as contrasted with—not necessarily opposed to—the rights of society.

Older tradition emphasized telling people what to do or what not to do, not just for their own protection but to develop an orderly society. Now the emphasis is on society protecting the rights of the individual, in some cases with what might be interpreted as resulting in inhibiting the progress of society. The result is a conflict between what can be called a "quality of life" ethic, as opposed to a "sanctity of life" ethic. One objective is to increase the quality of life of society. The other, often in direct conflict, is to protect the sanctity of each individual, whether or not that individual meets the definition of a "quality" life.

This fundamental consideration is at the heart of practically every ethical problem I have discussed. Abortion is a method for controlling population, and thus benefits society. It can be used to eliminate those who might be a cost to society because of physical handicaps. On the side of the individual it is argued that the mother should have the right to decide whether or not a child will be born. Paradoxically, the right of the individual to be born is used as an argument in opposition to the right to have an abortion.

A single individual, taking part in a medical experiment, might be the means of accomplishing a major step forward in medicine benefitting thousands or millions, yet doing that experiment is not justified unless some good also accrues to the benefit of the individual subject, or unless the individual agrees to be part of the experiment.

The widespread use of birth control would go far toward relieving misery and suffering in the world. Yet, because of the rights of each individual, no hint of coercion can be permitted in carrying out the programs that would have obvious benefits for society. Even the right of an individual to refuse medical treatment, a right the patient always had, has now come to be accepted generally.

A more complicated situation is brought about by the potential of gene manipulation. Whatever gene changes are made will be passed on to progeny. Do we have the right to determine what an individual yet unborn will be? What are the rights of future humans?

In *The Abolition of Man* C.S. Lewis says: "In reality, of course, if any one age really attains by eugenics and scientific education the power to make its descendants what it pleases, all men who live after it are the patients of that power. They are weaker, not stronger: for though we may have put wonderful machines in their hands, we have preordained how they are to use them."[1]

It is natural that medicine and the various sciences related to medicine should be the recipients of criticism as the result of the changing attitudes of society. The individual physician's primary responsibility has always been the individual patient. But the image of medicine for the public was that of a great discipline protecting society. In my own youth, the quarantine sign on some doors seemed to be visible constantly—measles, chicken pox, diphtheria, scarlet fever, mumps. The varicolored signs, keeping those outside out and those inside in, were not tacked to the door for the protection of the patient. They were there to protect society. The clinical trials carried out on prisoners in order to find a drug for the cure or prevention of malaria during World War II were of no benefit to the prisoners. The potential benefit was to society.

Now, with the new consciousness of the right of the individual, not only the benefit to society is important but also both the risk and the benefit to the individual must be considered. Thus, since medicine deals with individual human beings it is natural that it should be the subject of intense scrutiny when the rights of these same individuals are in question.

Every medical experiment carries some risk. Most medical experiments have some potential good, otherwise they would not be performed. While the correctness of these statements is generally appreciated, the ability to judge both risk and benefit has resulted in violent disagreements over many medical procedures. The physical risk is, in most cases, measurable. However, more and more people are questioning the ethical and moral benefits and the ethical and moral risks. And here the facts are not obvious and not measurable to everyone in the same degree. A decision on what is good or what is bad is based on the values of the individual making the decision. And there is a vast difference of opinion existing in definitions of good and bad.

It is easy to become seduced by the value of helping thousands of individuals. If a few get hurt on the way, so be it. The theological principle of proportionate good, that is, that the good consequences outweigh the bad act, is often used as an argument for justifying activities of this kind. However, if the act itself is considered unethical or immoral the end would not justify the means.

From this it is concluded that the goodness or badness of an experiment is not determined by the results obtained. The mere obtaining of valuable data does not make an experiment ethical or moral. It must be ethical from the beginning. Much valuable information was obtained from the human experimentation carried out in the German concentration camps. But the results did not justify the experiments. They were evil in themselves, even though benefit was derived from them.

If an ethical conclusion is measurable only against standards of values held by those drawing the conclusion, and if they differ in their standards, then their judgments of right or wrong will conflict.

Do scientists and ethicists, among themselves, have different value systems? Consider the statement of Leon R. Kass:

> Procreation is not simply an activity of the rational will. It is a more complete human activity precisely because it engages us bodily and spiritually, as well as rationally. Is there, perhaps, some wisdom in that mystery of nature which joins the pleasure of sex, the communication of love, and the desire for children in the very activity by which we continue the chain of human existence? Is not biological parenthood a built-in "mechanism," selected because it fosters and supports in parents an adequate concern for and commitment to their children? Would not the laboratory production of human beings no longer be human procreation? Could it keep human parenthood human?[2]

Now, contrast the position of Joseph Fletcher:

> It seems to me that laboratory reproduction is radically human compared to conception by ordinary heterosexual intercourse. It is willed, chosen, purposed, and controlled, and surely these are among the traits that distinguish Homo sapiens from others in the animal world. . . . Coital reproduction is therefore less human than laboratory reproduction. . . .[3]

It should not be surprising that scientists disagree on either the ethics or the benefits of an experiment. Values are usually not formed as the result of rational decisions based on data. So scientists as a group should not be expected to have any greater uniformity of agreement than any other group. However, the disagreements on things related to science have developed in the general public a sense of uneasiness. If scientists can't agree on science, who can?

It is but one of the results of the scientists' fall from grace that the public now knows that every scientist does not know everything, although some scientists are self-assumed experts in all fields. It should be encouraging to realize that the public is now becoming more critical in its judgment of who is capable and who is not capable of arriving at an informed decision in science. The fact that the organic chemist arrives at a ridiculous conclusion in interpreting epidemiological data is surprising only in the fact that the chemist believes that he or she is qualified in a foreign field. The same situation prevails across the numerous scientific disciplines. The information resulting from scientific research is increasing at such a rate that scientists have difficulty keeping abreast of their own specialty disciplines, without attempting to become an expert in others.

Dr. Jay Leslie Glick of Associated Biometrics Systems, Inc. has illustrated the problem. He says, "For the past two centuries our collective capacity to discover and assemble information in the form of scientific papers has been doubling every fifteen years . . . approximately 90 percent of all the world's scientists who ever existed are publishing in our lifetime more than 90 percent of all the scientific papers that have ever been written."[4] He describes as an example of the difficulty of keeping up to date the 1973 meeting of the Federation of American Societies for Experimental Biology where "during a four-and-a-half day period 21,544 registrants attended 453 sessions of which a maximum of 60 were being conducted simultaneously. In 410 of these sessions, 4,578 papers were presented with a 15 minute maximum on the amount of time permitted per paper."[5]

We have gone through the past cycle where anything that resulted from science was good, to the present where every change resulting from science is viewed with suspicion unless it can be proven to be good. In many of the present problems "good" includes, as a major component, "safe."

Most people now realize that it is impossible to prove a negative. In our consideration the negative is that some action will never prove harmful to someone under some conditions. By definition, then, nothing can be proven to be safe within the lifetimes of the users. This applies to drugs, to pesticides, to atomic energy plants, to the production of automobiles. Scientists are now being held responsible for any unforeseen results of any of their experiments.

I believe scientists should be held accountable, but with some major qualifications attached to that accountability. It is unrealistic to suppose that the scientists could ever be totally responsible for the manner in which the results of their work are used. The only situation in which a scientist could exercise complete control would be if all the scientists in the world united to refuse to do certain work, an unlikely possibility to say the least. Individual scientists have followed their own consciences in avoiding such projects as atomic energy, chemical warfare, and, more recently, recombinant DNA. But, where work has been done and information has been obtained, the scientist is only one of the elements in society contributing to a decision as to its use.

Because of the new awareness of the potential for harm from any piece of research the scientist does have a special responsibility. But because any research program might result in harm does not mean that every one will. Therefore, it is the responsibility of the scientist to try not only to postulate what effects might result, but also to estimate the chance that they will result.

In many areas of industry and government, approval to proceed with a building, a new technical process, a road, depends on the filing of an environmental impact statement. The objective is to predict, not just obvious results of the completion of the project, but also those peripheral consequences to, for example, the atmosphere, housing, employment, traffic patterns, etc. A similar assessment might be required for the introduction of techniques based on genetic manipulation, human experimentation, and even the use of drugs. For example, the introduction of insulin made it possible for many diabetics to live a much more nearly normal life. This is obviously a benefit. At the same time insulin made it possible for diabetics to live to the age of reproduction and to be responsible for the birth of children, some of whom could also be diabetics. In the early days before insulin, the diabetic population was controlled by natural selection. The sufferer usually did not live long enough to reproduce. I don't suppose anyone would seriously suggest withholding insulin in order to decrease the number of diabetics in the population. Nevertheless, the use of insulin could be considered to have this bad effect, in addition to its benefits.

The introduction of the birth control pill caused both a medical and a social revolution. It put into the hands of individuals total means for controlling population. Without hard data, the pill has been accused of encouraging promiscuity. Whether or not this accusation is just, it did allow the practitioner of promiscuity to do so with peace of mind. Now, long-range side effects are appearing that make it evident that there are significant risks that must be taken to obtain the great benefits associated with family planning and population control.

It is ironic that we got into trouble with science and technology only when we started to use its results. In the ancient "natural philosophy," the prime objective was to understand nature. For centuries major scientific errors were propagated simply because experiments were not run. Since control of nature was not even thought of, technical data were not necessary, and were not obtained. With the desire now to control, rather than just understand nature, it is necessary to experiment more and more, and to use the results of those experiments in some practical way. As our ability to experiment becomes greater, as techniques for manipulating the human organism become more generalized, as we approach the place where every natural function of the body can be imitated by a drug, a device, or a machine, it is inevitable that the results of science will begin to encroach on the laws of nature.

In our very efforts to become more human, we have, in some ways, dehumanized ourselves. No one could quarrel with the desire to save the life of an individual. Through the ages the medical profession has been dedicated to using all means available to prevent death, to prolong life. Now, our technological ability casts doubt on the desirability of that ethic. For example, it is now possible to resuscitate people who have technically drowned, who not too many years ago would have died. But in some cases a terrible price is paid. With the elimination of blood flow to the brain the patient can still be physically resuscitated, but has suffered severe brain damage, and is then doomed to exist as a vegetable for the rest of his or her life.

A particularly glaring example of what I consider to be an unnecessary and unjustified attempt to preserve life occurred in June of 1977. Siamese twins had been born to a family in Southern Italy. The body or bodies consisted of two heads, four arms and four shoulders, but only one torso and two legs. A team of doctors in Washington studied the twins for two and a half weeks, learning all they could about the anatomical composition of the bodies. After consulting lawyers and ethicists, and after obtaining the consent of the parents, an operation was performed with the stated objective of saving the lives of both of the twins.

It is difficult to rationalize such a decision. Regardless of how you visualize the resulting two bodies, there just aren't enough parts for two human beings. The operation was probably performed brilliantly, but it seems to me that, in spite of the good objective, it violated all the ethical principles that should have been observed in allowing the twins to die a merciful death. The attempt at separation of one and one-half bodies into two separate human beings is an example of an indulgence of our technological skills. The enormous effort that went into this operation was an unjust use of valuable resources and so, in my opinion, made the entire procedure unethical, totally apart from the extraordinary suffering caused to the patients. Because a means was available, it was used.

Somewhere there must be a middle ground between those who say that lives must be preserved at any cost, that living in any way is better than dying, and those who say life, short of perfection, is not life.

The use made of resources raises a whole host of other ethical problems. How should money, highly trained scientists, equipment, physical facilities, be distributed in a just and equitable manner? The problem is not just how such priorities are to be set among medical applications, but how do we decide that medicine is more or less important than housing or education or other necessities for a good life?

Such priorities are being set every day, for it is obvious that every need, real or perceived, cannot be met. In one state a bill was introduced that would have resulted in an expenditure of many millions of dollars to relieve a serious situation in the state mental institutions. The bill was rejected because legislators felt that there were better ways to spend the funds. They pointed out that the same amount of money would buy hot lunches for all the school children in the state, and could provide job training for many. Alternatively, health care services could be provided for normal or near-normal children, and pregnant women. It was stated that much mental retardation could be eliminated through prenatal diagnosis of Down's Syndrome at $200 per case, compared with $60,000 for each institutionalized child.

In 1972 about 40 percent of all dialysis patients were treated at home. In 1977 the government estimated that the percentage had dropped to 10–13 percent. By 1979 the expenditure is expected to rise to about $1 billion annually. Dialysis at home averages $6,000 to $8,000 annually, compared to $24,000 in a dialysis center, and $30,000 in a hospital. In spite of the convenience of home treatment the number of such patients is steadily decreasing, and institutionalized treatment is increasing. The reason is strictly financial. Patients treated at home are required to purchase their own supplies. In the hospital, the combination of government financing and insurance pays 100 percent of the cost. To reverse the trend of the increasing cost of the dialysis program, Congress passed in May of 1978 a bill that would approve the payment of the cost of supplies for home dialysis, and thus eliminate the financial differentiation to patients between the home and institutional dialysis. (A heart transplant, by comparison, costs between $20,000 and $50,000.)

Are there other ways the money could be distributed with greater justice? We might even ask, "Should we try to prevent death by curing cancer, by eliminating heart disease, and other major killers?" What purpose does it serve? It would not eliminate death. It would only mean the end of life would be caused by some other affliction. It probably would not extend the life span of individuals more than a couple of years if cancer and heart disease were both eliminated as causes of death.

As a matter of cold analysis, it would probably be better to expend the resources—money, technical ability, human energy, on prenatal care, on improving the lives of all rather than on trying to prevent the unpreventable for some. What would be the effect on our standards if we were to assume that death was inevitable, that it was useless to try to cure other than infectious diseases? Assume that the same effort would go into increasing the standard of health living for all as now goes into attempting to prevent death. The most obvious effect would be that some people would die earlier than they would have with life-prolonging treatment. A true analysis would ask at what cost—in money, and in mental and physical suffering—does the individual live out those few extra years?

If such a posture could be considered possible, the consideration is endless. If there were no heart diseases and no cancer there would be other conditions taking their place, causing death in the same inexorable fashion. Is it futile to attack all causes of death, particularly when such a direction means giving prevention of death priority over some actions that could result in a better life? Logic answers yes. The inevitable conclusion is that no efforts should go into finding cures for diseases that cause death, that expenditures of money and time should be directed to making more meaningful the lives that are lived. But having presented the problem I find it impossible to say we should still not try to find a prevention or cure for cancer and other killer diseases.

Hard choices? There are no easy choices left.

I shall not say that science has solved old problems. It is more accurate to say that science has made it possible to do many new things. Some of these things result in the solution of problems but at the same time they can cause equally severe new problems. Each advance gives us more freedom, but with each new freedom we become more the prisoners of our acts.

I have found it difficult to formulate and express ethical judgments on scientific questions. Probably it has been made more difficult because of my forty years of involvement in research, and in the necessity to change some of my early attitudes regarding the relationship or results of technology and the interest of society in the use of that technology. It is difficult to separate the rights of the individual from the rights of society. It seems right to ask that every individual be guaranteed the requirements for a decent existence before being allowed to be born. It helps both those born into society, as well as the society into which they are born. Yet I think it is the height of elitist arrogance to define that existence as that which we require for our own satisfaction, even our own indulgence. We decide for people unborn what we think they would want and, when we are criticized by those more closely associated with the problem than we are, we accuse them of being ungrateful for all the help we have given them.

How then does one rationalize the dilemma of the rights of the individual when the exercise of such rights either does not benefit society or might actually harm society? My response would be that rationalization is not necessary or possible. The primacy of individual rights is either accepted or not accepted.

A better question to ask might be how does one make the results of the exercise of individual rights coincide with the needs of society. It is not an easy question to answer, because the needs of society change. Needs are also determined by values, and values change. In the early days of our own society the country was based on agriculture. The many small farms, operated by individual families, required hands to do the work. Large families were needed, and the more male children the better. Because of high infant mortality it was assumed that more children would be born than would reach the age where they could contribute their efforts to the family farm operation. Thus, there was a strong stimulus to the value that said a large family was good. However, two things happened in a short space of time to change society's needs. As farms became absorbed into huge agricultural complexes, the need for the large nuclear family was eliminated. In addition, the infant mortality rate declined dramatically, with no decrease in birth rate, so that what was once a requirement for existence now became a burden to society.

The problem of changing population demands and pressures is typical of the dilemmas caused by increasing knowledge and advances in science and technology. Although the needs and values change, the rights of the individual and of society did not change. An obvious and direct solution to this particular problem would be to impose some legal restriction on the right to have children, or to impose economic sanctions on those who have more than whatever number is deemed permissible by the state. The number usually spoken of now is whatever is required for a "zero growth society"—each couple having two children to replace themselves. Such restrictions have been suggested.

In India a program of compulsory sterilization was instituted, but because of violent popular opposition it was soon dropped. If the rights of the individual are to be preserved, the only solution must be a campaign to convince the individual that each does have some responsibility to society; this will result in voluntary limitations to family size.

The same principle is at the root of the conflict as to whether or not recombinant DNA research is to be conducted. However, in this situation, other considerations are involved. Scientists claim the right to do whatever research they want. Any restrictions put on the kind of research done is an infringement on the scientist's rights as an individual. However, I do not consider the right of a scientist to do research to be comparable to the right of parents to bring children into the world. The right to parenthood is a fundamental right residing in the individual because of his or her humanness. An extreme analogy would be the conflict between a scientist who claims the right to experiment upon a patient, and the right of the patient to accept or reject such experimentation. Obviously, the right of the patient must prevail. So the right of the individual scientist does not assume the right to act without restriction.

Scientists have a vital role to play in solving each of these dilemmas. They have been accused of causing them. Whether or not this is entirely true, they do have a responsibility to aid in eliminating them. Their role is unique because of their knowledge. Unfortunately, it is a responsibility that scientists have not been accepting.

The freedom of expression of the poet, the dramatic visualizations of the artist, the impassioned pleas of the lawyer have all been accepted as characteristic of these professions. But the scientist has always been thought of as one who, by the very nature of his or her work, must react to any situation with a total lack of emotion. This emotionless approach has been equated to an objective approach, and as a corollary, any emotional involvement in a problem automatically eliminates the possibility of objectivity. This is nonsense. The time is long past, if in fact it ever existed, when the scientist could say, "Here are the results of my efforts. Use these results as you will, for good or evil. I have done my part. The consequences of these acts are now your responsibility." The time has come when scientists must assume responsibility not only for their own work but for the eventual uses of that work.

It is right sometimes that a person should react with the passion of the artist rather than with the discipline of the scientist. Scientists owe themselves and their profession only this code: that they speak with authority, that they speak with honesty, and that they speak with objectivity. However, having arrived at a conclusion as a result of objective analysis, it is right that they speak also with passion and emotion, for they are only doing what they must do in the face of situations that tend not only to injure science but to prevent each scientist from contributing to the fullest for the progress of the whole human race.

Too often scientists have not been fulfilling their responsibility to educate. They are by nature usually averse to adversary proceedings, to conflict, to confrontation. Rather than entering into public disagreement they simply leave the field, usually to those less competent to interpret their results and to lobby for their objectives. The result has been predictable. In a world where sensation is the order of the day, it is no surprise that scientific work must be sensationalized to draw attention. It is no surprise that much of the sensationalism comes from a prediction of evil consequences. The scientist is at a disadvantage when debating possibilities, rather than facts. If he or she cannot promise that an evil consequence will not happen, there are many outside of science who are willing to say that it will happen.

And so we see the situation that exists today—a public demanding that activities be risk-free, the scientist unable and unwilling to say that anything is risk-free, the public then concluding that, if it can't be guaranteed risk-free, it must be dangerous. And so the popular press is eager to report the 0.1 percent side effects of a drug, rather than the 99.9 percent cures. Or it will report the speculation that a product might cause cancer, rather than the fact that the product has saved millions of lives.

Even government agencies, responsible for protecting the health of the nation, have been guilty of overreacting to predictions of catastrophe founded on little or no data. In 1973, based on a report from a single physician, the United States Consumer Products Safety Division announced that it was banning spray adhesives because of the possibility that they caused birth defects. Six months later the ban was lifted after further studies showed that the spray was safe. However, during that six months, the Commission's action received wide publicity. In February 1976 an article appeared in *Science* reporting the results of a survey that had been taken of 182 genetic counseling centers, and pointing out one tragic consequence of the reckless and irresponsible actions taken by the Consumer Products Safety Commission. At least eight women who had been exposed to the spray adhesive, and had also been exposed to the publicity of the Agency, elected to have abortions without undergoing tests that would have told them whether or not the fetus was defective. Other women asked for chromosomal screening as the result of this fear.

The *Science* article pointed out that some effort should be made to distinguish *suspicion* of toxicity from *evidence* of toxicity.

The opposite reaction, namely that of refusing to approve for use a product accused of causing side effects, is not uncommon.

One cannot quarrel with the policy that says, "If we err, we will err on the side of conservatism." But when conservatism is interpreted as refraining from any positive action, then I do quarrel, not with the policy, but with the inaction. It is safer to do nothing, not necessarily safer medically, but certainly safer politically. If you do nothing you cannot be accused of making a mistake. However, there will be no positive benefits either. It is easy to measure the benefits and the costs of something that is done. However, there is no way to measure what has been lost in benefits by something that is not done.

Uncertainty is the characteristic of every scientific endeavor. The uncertainty is magnified when the reaction of an individual human being is one of the factors involved in evaluation of an experiment. For example, if a drug is to be tested on one million human beings, the first 999,999 individuals could react identically, but there would be no assurance that the one millionth would not react differently. There would, obviously, be the very strong possibility that the reaction would be the same, but it is just such uncertainty that makes it impossible to argue against the statement that some ill effect might take place, regardless of how remote the possibility.

Among the problems that I have discussed, that of recombinant DNA is the best example of a scientific activity that can have tremendous benefit for society and, at the same time, lends itself to speculation as to all types of science fictional horrors occurring in the future. The speculated horrors cannot be guaranteed against happening. But it can be said that the possibility of catastrophe is infinitesimally small.

Inevitably, when the question relating to dangerous research that might produce dangerous results is discussed, the question of control of research is also raised. Science is too important to be left to scientists. The public must be involved in decisions that might affect them. Therefore, the public must control science. The answers to several questions would give at least some guidance for future planning. First, does the public have the right to control or regulate scientific research and technology? If we agree that anything affecting the life of the public should be under public control, then certainly science should be in that category. Science, whether one likes to recognize it or not, has become essential to our existence and will continue to be so. The more important it is, the greater the necessity and justification for control. Science is no longer a private pursuit of scientists. It is now, or should be, a servant of the public.

The second question that must be asked is whether or not the public has the ability to make the decisions necessary to regulate science. The primary objection that scientists have to any form of regulation is that no one but a scientist can understand science. The statement as it stands is true. But we must distinguish between the mechanisms of science, that is, the methods used to solve a problem, and the effects that result from the solution of the problem. It is really the results of science and technology that affect the public. Therefore it is not necessary for those making decisions to have the same degree of specialization as the practicing scientist in order to decide whether a result is good or bad. Very few, if any, legislators understand the science behind the atomic bomb. Yet decisions affecting atomic energy can and must be made. It is not even necessary to understand the working of an internal combustion engine in order to establish pollution controls.

The words of the Cambridge Citizens' Committee that considered whether or not to allow Harvard to conduct recombinant DNA research in its city limits are worth noting:

> While we should not fear to increase our knowledge of the world . . . we citizens must insist that in the pursuit of knowledge appropriate safeguards be observed by institutions undertaking research. Knowledge, whether for its own sake or for its potential benefits to humankind, cannot serve as a justification for introducing risks to the public unless an informed citizenry is willing to accept those risks. . . . We wish to express our sincere belief that a predominantly lay citizen group can face a technical scientific matter of general and deep public concern, educate itself appropriately to the task, and reach a fair decision.[6]

Events taking place since the publicity attendant on recombinant DNA research have made me apprehensive concerning the approach the public will take in using what it perceives to be its newfound power, not over science, but over the scientists. Any issue that can be inflated to the point where results of activities can be said to affect not only every living individual but also future generations is a magnet to publicity-seekers. And the same disagreeable tones of anti-business, anti-science, even anti-intellectualism that I thought had died deserved deaths are being heard again. At a meeting sponsored by the prestigious National Academy of Sciences and held in Washington to discuss recombinant DNA research, the participants were greeted by a group of objectors singing, "We shall not be cloned" to the tune of "We Shall Overcome." A scientist is especially frustrated by such a reaction since recombinant DNA is not related to cloning in any way.

At the same meeting, the leader of a Washington-based citizens group stated that the DNA work would lead to genetic engineering, and that his group had the means to resist it. He said: "A storm of public outrage is coming, and it won't be gentlemanly." What "not gentlemanly" means I don't exactly know. I can only conclude that the term hints at violence as a substitute for rational discussion or an attempt at understanding.

The same group pointed out, in a not very subtle fashion and with an implication that everything was not honest, that the expenses of the National Academy meeting were partly underwritten by eight pharmaceutical companies who were interested in recombinant DNA research. If one begins with an extremist position, I suppose it is difficult to recognize objectivity in anyone else. It probably never occurred to those casting suspicion on the industrial scientists that the scientists were as interested—possibly more so—than was anyone else in arriving at the proper conclusion both as it concerns science and as it concerns ethics.

It irritates me that scientists working in industry are assumed automatically to lack objectivity about their work. No such assumption is made about academic or government or citizen's group scientists. Yet, I submit that the industrial scientists, particularly in the pharmaceutical industry, have been faced with more ethical decisions and have made them more justly than any other group of scientists. Anyone who doesn't believe that is ignorant not only of what research has been done, but of what research could have been done. Obviously, pharmaceutical scientists are interested in the commercial application of recombinant DNA techniques. I have already outlined some practical potentials. But they are also as aware of the ethical demands placed on them as are those whose voices are constantly heard warning of "commercial vested interests."

It seems to me that there is as much reason to start demanding that the public act responsibly as there is to demand responsibility from scientists. It is sometimes difficult to interpret as irresponsible an action that has as its ostensible objective the protecting of the public from a catastrophe. However, even with such a laudable objective, I think it is irresponsible to arrive at decisions based on a prejudgment that evidence is biased because it has been collected by a particular group of scientists, whether they be in industry or one of the academic fraternities, or even worse, to ignore evidence completely because it might result in changing a conclusion.

The question at the heart of much of the disagreement is not just what research shall be done but who will control it. Much of the public discontent is based on the simple fact that they weren't consulted. This has been stated rather bluntly by some individuals involved. "You arrogant scientists think you can do anything you want. Now you know you have to go through us." While scientists have been preoccupied with what they are doing, the public is concerned with how the results will be used. I believe this is a logical starting point for defending the responsibilities of each segment of society. Scientists must be free to decide how they do research, who is capable of doing research, and the quality of that research. It is only when research results in a consequence to society that the public must be part of the decision-making process.

The difficulty, however, lies in the fact that the public must still depend on the scientist for much of the information it uses to form a judgment. For example, only the scientist can give an estimation of the adequacy of NIH guidelines as they relate to lab safety. A scientific estimate can be given of the chances for an accident, and the consequences of that accident. The public must then weigh the evidence and decide whether, in view of the consequences, the risk is worth taking. The factor complicating arrival at a decision on the safety of DNA research has been the conflicting views presented by scientists. I can only suggest that, just as the expertness of the witness is judged in courtroom evidence, so also the expertness of the scientist should be judged. The Nobel laureate whose voice is heard in any scientific discipline may not be nearly as expert as the less famous individual who has spent a lifetime learning all there is to know about the specialty under discussion.

Please understand that I do not exclude the scientist from a part in this decision. There are not two classes of people—scientists and the public. The scientist is part of the public. In addition, scientists have a tremendous specific responsibility. They are the ones who must educate the public to the consequences of the work, to the risks and to the benefits. To the extent that the scientists do an effective job in this, and to the extent that the public is responsive, the resulting decisions on controls will be responsive to human needs.

Too often scientists in the past have been reluctant to include non-scientists in the inner circle of their knowledge. Control is seen as an intrusion. Therefore the scientist believes that, if the public can be kept in ignorance of scientific work, there will be no control and therefore no intrusion.

I think the opposite is true. With the importance science has assumed in our lives, there will inevitably be controls. I would much prefer to see these controls result from intelligent opinion informed through the guidance of scientists, than to have controls based on insufficient data that might result in over-restriction caused by emotion.

I am not optimistic about scientists being able to make the necessary judgments to control research that might be dangerous to the public by self-limitations on their own work. The example of control of gene research is the only instance I know of where an attempt has been made to do this, and it is much too early to judge results. Even here, the scientists involved directly in this decision represent such a small percentage of those capable of doing the research, regardless of quality, that it is doubtful that their decisions and recommendations would be adopted worldwide. I fear that many scientists believe that their research is their private possession, that their knowledge is their private knowledge, and that they should be allowed to pursue whatever work they choose.

The suggestion that knowledge resulting from recombinant DNA, genetics, or any other research should not be obtained because it is too dangerous is repulsive to me. I do not think there is any knowledge too dangerous to possess.

A prominent scientist has said, "I would gladly trade off the knowledge of nuclear science presently possessed by human society for the opportunity to live without the anxiety of a nuclear holocaust or the rapid spread of radioactive material." I think that is a ridiculous attitude. To eliminate any knowledge or technique that might result in risk, regardless of the benefits, is a futile desire to return to the "never was" time of "natural" living. It is as much a fantasy as is the "never will be" time of responsibility-free societal paternalism or the discomfort-free age of technology.

Knowledge is what gives us freedom. And if we want freedom we must pay for it. Part of that payment is the risk that we must take. Knowledge also gives us power. In one sense, having power makes us less free, because power in science, like all other power, has the ability to possess its owner. Whether we are free or enslaved because of knowledge does not depend on what knowledge we have. It depends on how the knowledge is used.

It is no small thing, being a god. Gods have problems, too. Gods can only give humans tools for survival. The tools must be used effectively. But the gods must select proper tools.

When Epimetheus distributed to each living creature the powers and characteristics necessary for the preservation of its species he also gave us an example of what is necessary for survival. The birds were given the ability to fly away from danger, while their predators were earthbound. Some small, weak creatures were given the ability to burrow underground. Those relatively defenseless were made prolific, so that by sheer numbers some would survive. The big and the strong were made less prolific, so that they would not overrun the earth. Even those totally defenseless animals were given the gift of camouflage.

However, all the gifts of Epimetheus had been exhausted before he came to preparing humans for emergence "from within the earth into daylight." Humans had no equipment to survive in nature. They could not fly, they could not burrow, they could not move with great speed, they were not strong, they had no weapons, they had no camouflage. It was then that Prometheus came to the rescue with the gifts of skill and the tangible gift of fire. The ability to use fire was humanity's first venture into technology—the first venture that changed nature. But even in the myth it was recognized that this gift was not enough for survival, and could even be dangerous. Zeus, fearing that humanity would destroy itself, commissioned Hermes to give humans two other virtues—reverence for others, and a sense of justice. He even went so far as to say that "He who has no part in reverence and justice shall be put to death, for he is a plague to the state."

Our skills have enabled us to change the environment in which we live. We are now able to communicate almost instantly with people on the other side of the world. We can be transported to them in a matter of hours. We can free ourselves from the confines of earth. Now we are able to modify ourselves. And it is this ability that makes the advent of genetic manipulation both hoped for and feared. Is it any wonder that it is thought that skills alone might be dangerous to our survival?

For all practical purposes, skills are available to solve most of what we call the major problems of the world—controlling population growth, feeding the hungry, producing energy, increasing the health level, cleaning the environment, etc. Any additional technology required is simply a modification of what we now have, plus what could be obtained if we decide it is worthwhile to obtain it.

In the world determined by Epimetheus and Prometheus, the human being functions by reason, all other creatures by instinct. The results of our reasoning are determined by our value systems. The qualities demanded by Zeus—respect for others and a sense of justice—mean many different things to many different people.

The problems of our time are not caused by technology. They

are moral and political. What the world will be in the future will not be determined by sensational new scientific advances. It will depend on how well all of us can organize and agree on common objectives. More important, it will depend on how much we are willing to pay to reach these objectives.

We'd better get good at it.

epilogue

To the casual observer everything seemed to be going along smoothly at the plant. The ova collectors had done a good job of building up a group of donors with the proper genetic composition. Some donors had been given superovulators, resulting in the production of numerous ova every month. Efficiency over the last quarter had increased significantly.

With the new method developed in research the Quality Control Department was now able to detect chromosomal abnormalities immediately after fertilization in the test tube, so that it was no longer necessary to waste synthetic uterus space on embryos that would eventually be rejected. Early elimination of rejects also made it easier for the Distribution Department to plan for the priority delivery of fertilized eggs to the farm for implantation into the uterus of normal cows. Those fertilized eggs to be transferred from the test tube to the synthetic uterus did not present the same urgent problem, since it did not make too much difference whether or not their degree of development was carefully controlled.

Since the ovum fertilization farm had been converted to agricultural uses all the morale problems among employees had disappeared. They realized the tremendous benefits that could result from using this technique, and they were glad to be a part of such a wonderful development. Human embryo transplant had not been completely abandoned, but it was now used only in specific individual cases. The custom of eliminating the elderly and the unfit, too, was no longer accepted, and genetic modification was attempted only in those cases where physical or mental problems could be eliminated or prevented.

The trouble really started in what is now remembered as "The Time of the Questioning." It is also remembered as the time when people found out they were free to think. Education was questioned, government was questioned, religion was questioned, even science was questioned.

How could science be questioned? Science, the great benefactor that had given people a longer life with baubles for their comfort and convenience beyond their wildest dreams, that had even given them the power to modify not only individual human beings but whole species.

The questions did not begin during the time when the whole human raced looked on science with respect. It began only when some began to look on it with fear, a fear generated by the unknown, by the evil possibilities, and, with increasing frequency, by the use to which the results of science were put. The questions did cause trouble for science. Everything was questioned and, in the beginning, everything was feared. But with questioning also came understanding. As time went on the fears of the terrible potential of science for evil were not diminished, but there developed a realization that science might still be the great benefactor it was once thought to have been, if only its results could be directed.

With the ability of people to think and to question also came a realization of the responsibility that was now theirs. They recognized their dependence on each other, and they recognized the rights that each individual had. Decisions were made only after everyone had been fully informed of the consequences, and after everyone had had a chance to participate in the decisions. Mistakes were made, but they were their own mistakes. No longer would they be told what to think, what to eat, what to wear, what kind of cars to drive. The mistakes they made were the price they were paying for their freedom. They thought it was worth it.

The Humanity Modifiers, seeing the direction the nation was going decided that they could do nothing but join the parade. They, too, decided to seek advice, to give information before making a decision. They thought they should give the system a chance. It might even work. After all, it had never in all history been tried before.

notes

CHAPTER TWO

1. Edmond Cahn, *The Moral Decision* (Bloomington: Indiana University Press, 1955).
2. Garrett Hardin, "Living on a Lifeboat," *Bio-Science* 24 (1974): 561–68.
3. Paul Ehrlich, *The Population Bomb* (New York: Ballantine Books, 1971).
4. Amnon Goldworth, letter to the editor, "Aboard the Lifeboat Debate," *Hastings Center Report* 5, no. 2 (April 1975): 43.
5. Daniel Callahan, "Doing Well by Doing Good: Garrett Hardin's 'Lifeboat Ethic'," *Hastings Center Report* 4, no. 6 (December 1974): 1.
6. Alan F. Guttmacher, *Medical World News,* November 1974, p. 56.
7. John R. Cavanagh, ibid.
8. Charles Lowe and Colin Roberts, "Where Have All the Conceptions Gone?" *Lancet,* March 1, 1975, p. 498.
9. Kansas Session Laws of 1970, CH–378.
10. *Journal of the American Medical Association* 205 (1968):337–40.
11. Speech to the International College of Anaesthesiologists, November 24, 1957.
12. Ibid.
13. Ibid.
14. Act 3, sc. 1, line 104.

CHAPTER THREE

1. H. V. Aposhiam, *Perspectives in Biology in Medicine* 14 (1970): 98.

CHAPTER FOUR

1. Aristotle, *Politics,* VII. 16. 1335b20.
2. Bentley Glass, "Science: Endless Horizons or Golden Age?" *Science* 171 (1971):23.

3. Leon R. Kass, "The New Biology: What Price Relieving Man's Estate?" *Science* 174 (1971): 781.

4. René Dubos, *The Condon Lectures: The Cultural Roots and the Social Fruits of Science* (Eugene, Oregon: State Board of Higher Education, 1963).

5. René Dubos, quoted in Eliot Marshall, "Environment Groups Lose Friends in Effort to Control DNA Research," *Science* 202 (1978): 1265.

CHAPTER FIVE

1. Marc Lappé, James M. Gustafson, and Richard Roblin, "A Report from a Research Group on Ethical, Social, and Legal Issues in Genetic Counseling and Genetic Engineering of the Institute of Society, Ethics, and the Life Sciences," *New England Journal of Medicine* 286 (May 25, 1972): 1129–132.

2. Marc Lappé, "The Genetic Counselor: Responsible to Whom?" *Hastings Center Report* 1, no. 2 (April 1971): 6.

CHAPTER SIX

1. Sir Francis Galton, *Hereditary Genius* (Cleveland: World Publishing, 1962).

2. Charles Darwin, *The Descent of Man and Selection in Relation to Sex* (New York: Appleton, 1896).

3. Margaret Sanger, *Woman and the New Race,* quoted by Jon Beckwith, *Annals of the New York Academy of Sciences* 265 (1976): 47.

4. Hearings before the House Committee on Immigration and Naturalization, 1924.

5. Theodosius Dobzhansky, "Man and Natural Selection," *American Scientist* 49 (1961): 285–99.

6. Joseph Fletcher, "Ethical Aspects of Genetic Control," *New England Journal of Medicine* 285 (1971): 776–83.

7. Bentley Glass, "Human Heredity and Ethical Problems," *Perspectives in Biology and Medicine* 15 (1972): 237–53.

CHAPTER SEVEN

1. Edward T. Tyler, "The Clinical Use of Frozen Semen Banks," *Fertility and Sterility* 24 (May 1973): 413; J. K. Sherman, "Long Term Cryopreservation of Motility and Fertility of Human Spermatozoa," *Cryobiology* 9 (1972): 332.

2. J. K. Sherman, "Synopsis of the Use of Frozen Semen Since 1964: State of the Art of Human Semen Banking," *Fertility and Sterility* 24 (May 1973): 397.

3. C. F. Westoff and R. R. Rindfuss, "Sex Pre-Selection in the United States: Some Implications," *Science* 184 (1974): 633–36.

CHAPTER NINE

1. Arthur L. Herbst et al., "Age–Incidence and Risk of Diethyl Stilbestrol-Related Clear Cell Adenocarcinoma of the Vagina and Cervix," *American Journal of Obstetrics and Gynecology* 128, no. 1 (May 1977): 43.

2. *Report of the National Commission for the Protection of Human Subjects of Biomedical and Behavioral Research* (March 1977).

3. U.S. v. Karl Brandt et al., Trials of War Criminals Before Nuremberg Military Tribunals Under Control Counsel Number 10 (October 1946–April 1949).

4. H. J. Muller, "Should We Weaken or Strengthen Our Genetic Heritage?" *Daedalus* 90 (1961): 445.

CHAPTER TEN

1. C. S. Lewis, *The Abolition of Man* (New York: Macmillan, 1965).

2. Leon Kass, "The New Biology: What Price Relieving Man's Estate?" *Science* 174 (1971): 779–88.

3. Joseph Fletcher, "Ethical Aspects of Genetic Control," *New England Journal of Medicine* 285 (1971): 776–83.

4. Jay Leslie Glick, "Ethical and Scientific Issues Posed by Human Uses of Molecular Genetics," *Annals of the New York Academy of Sciences* 265 (1976): 182.

5. Ibid.

6. *Chemical and Engineering News* 24 (January 1977): 6.

index

Abolition of Man, 150
Abortion, 10, 14–15, 17, 20, 56, 75, 100, 115, 126–28, 150
Achondroplasia, 48
Adenosis, 131
Aeson, 116
Aging, 24
Alcibiades, 53
Alfidi, Ralph, 145
American Association for the Advancement of Science, 56
Amniocentesis, 54–57, 98, 100; determination of sex, 55; determination of age, 55; determination of weight, 55; determination of lung development, 55;
Analgesics, 3, 5, 146
Anatomical Gift Act, 23
Anesthetics, 4
Antibiotics, 3–4, 7, 65
Anticoagulants, 144–45
Antihistamine, 3, 5
Aposhiam, H. V., 51
Aristotle, 40, 53–54, 101, 147
Arkansas Supreme Court, 21
Artificial insemination, 98–100; in agriculture, 99
Artificial uterus, 104, 108
Asilomar conference, 66–67, 71
Assembly of Virginia, 78
Augustine, Saint, 11
Autosomal disease, 47
Autosome, 43

Bacteriophage lambda, 60
Barbiturates, 2, 6
Becher, J. J., 3
Beckwith, John, 66
Behavior control, 119–20
Bevis, C. W., 55
Biomedical engineering, 115
Birth control pill, 154
Birth defects, 161; in artificial insemination, 99–100
Birth marks, 45
Blastocyst, 59
Blind study, 145
Blood transfusion, 115, 117–18
Bodmer, W. F., 88
Body bank, 118
Boulding, Kenneth, 12
Boyer, Herbert W., 63
Brain death, 23, 26; law, 22
Brown, Lesley, 104
Brown, Louise, 105–106
Browne, Thomas, 30
Butler, Justice, 91

Callahan, Daniel, 13
Camera della Segnatura, 53
Cancer, 5, 65, 79, 131, 133, 146, 157, 161
Catholic Church, 6, 106, 110, 118
Cavalli-Sforza, L. L., 88
Cavanaugh, John R., 17
Cell division, 42
Cell fusion, 70
Chemists' Creed, 3
Children, research subjects, 141
Chimera, 59
Choriocarcinoma, 5
Chromosomes, 39, 42–43, 45–46, 88

Cleft lip, 49
Cleft palate, 48–49
Clinical trial, drugs, 125
Cloning, 98, 112–113
Club foot, 48–49
Cockroach, 41
Commission on Hereditary Disorders, 82
Consanguineous marriages, 88
Contraceptive, 3, 6, 7, 109–11, 154
Control of research, 162, 164–66
Cooley's syndrome, 49
Cornea transplants, 118
Cosmic rays, 43
Crick, F. H. C., 41
Cystic fibrosis, 45, 48
Cystinurea, 48

Darwin, Charles, 39–40, 89
Death, 14, 19–20, 22–23, 28, 33
Delgado, José, 123
Del Zio, John and Doris, 107
Deoxyribonucleic acid (see DNA)
Department of HEW, 55
DES (see stilbestrol)
Dialysis, 156
Diethylstilbestrol (see stilbestrol)
Diogenes, 53
District of Columbia screening, 78, 81
DNA, 6, 40–41, 59–63, 65–68, 70–71, 73, 112, 165
Dobzhansky, Theodosius, 91–92
laevo DOPA, 47
Down's syndrome, 46, 51, 55–57, 80, 85, 156
Dubos, René, 71–72
Dying, 24

Edelin, Kenneth, 15
Education for All Handicapped Children Act, 95
Edwards, Robert G., 104–105
Ehrlich, Paul, 12, 72
Electrostimulation, 121–22
Embryo transplant, 104, 106
Environment, effect on genetic variation, 49; factor in intelligence, 93; effect on character, 94; risk free, 70
Epicurus, 53
Epimetheus, 166–67
E. coli, 60, 63–65, 67, 69, 71
Esposito, Elaine, 37
Estrogen, 6
Ethnic diseases, 48–49
Euclid, 53
Eugenics, 89, 93–94, 97, 150
Eukaryote, 59–60, 69
Euthanasia, 10, 35–38
Euthenics, 95–96
Evolution, 40–41, 65–66
Extra fingers and toes, 49
Extraordinary means, 30, 34

Federal Register, 71
Feeblemindedness, tests on immigrants, 91
Fertility, 111
Fertilization, 18–19, 102, 130; *in vitro,* 98, 102–104, 106–107, 111, 114; in agriculture, 103; abnormal effects, 108
Fetal experimentation, 14, 125, 129, 136
Fetus, personhood, 129; viability, 129
Fletcher, Joseph, 92, 152
Fox, Allen S., 59
Fraenkel-Conrat, Heinz, 58
Freedman, Benjamin, 140
Friends of the Earth, 72
Fuchs, F., 55
Fulton, James, 120

G6PD, 58
Galactosemia, 58, 75; effect of environment, 49
Galton, Francis, 89, 93–94
Genentech, 64
Gene pool, 114
Genes, 39, 41–44, 46, 50, 57, 60–61, 63, 65–66, 70, 87–89
Genetic control, 95
Genetic counseling, 83–85, 95, 161
Genetic defects, 74; determination of, 74–75

Genetic diseases, 39, 41–42, 45, 50–51, 57–58, 75, 82
Genetic engineering, 163; industrial use, 64
Genetic factors in intelligence, 92–93
Genetic manipulation, 67, 98, 150, 154; in agriculture, 65, 95
Genetic screening, 44, 54, 56, 64, 74–75, 78–79, 83, 95
Genetics, effect of chemicals, 62
Genetics of Human Population, 88
Genocide, 74, 76, 91
Genotype, 97
George III, 39
Gerald, Park, 76
Gillium, Richard, 49
Glass, Bentley, 54, 92
Gleitman, Jerry, 84
Glick, Jay Leslie, 153
Goldworth, Amnon, 13
Goodlin, Robert, 108
Goodman, Howard M., 63
Gray, Cardinal, 106
Gurdon, J. B., 112
Guthrie, Robert, 81
Guttmacher, Alan F., 17

Haemerli, Urs Peter, 23
Handler, Philip, 12
Hardin, Garrett, 12
Häring, Bernard, 106
Harris, Augustine, 106
Harris, Henry, 59
Harvard Medical School, Committee to Examine Definition of Brain Death, 26
Haymarket Riot, 89
Heart disease, 157
Heart transplant, 117
Heinonon, O. P., 133
Helligers, André E., 18
Hemoglobin, 39, 43
Hemoglobin-S., 47, 78
Hemophilia, 50, 102
Hepatitus, 141
Heracletus, 53
Herbst, Arthur L., 132
Heredity, unit of, 41
Hermes, 167
Hess, W. R., 122
Heterozygous, 43
Hitler, Adolph, 94
Holmes, Justice, 91
Holmes, US. vs., 11
Home, Everard, 99
Homocystinurea, 48
Homozygous, 43
House Committee on Immigration and Naturalization, 90
Hua T'o, 117
Human Experimentation, 143, 147, 151, 154
Human life, 17
Humanness, 11, 147
Hume, David, 118
Huntington's disease, 46–47, 80
Hurley, Mark, 106
Hybrid vigor, 114
Hypertension, 49

Ideogram, 46
Illinois Domestic Relations Act, 78
Illinois Supreme Court, 84
Immigration Restriction Act, 90
Implantation, 19, 103
Inborn errors of metabolism, 48
Infertility, 100–111
Informed consent, 121, 135, 138, 143–45, 148
Institute of Society, Ethics, and the Life Sciences, 13, 83
Insulin, 4, 63–64, 154

Jason, 116
Jenner, Edward, 142
Julius II, Pope, 53

Kahn, Edmund, 12
Kansas death definition, 25–26
Karow, Jr., Armand, 99–100
Karyotyping, 45
Kass, Leon R., 57, 152
Khorana, Har Gobind, 60
Kidney transplant, 118
Klinefelter's syndrome, 76–77, 80
Kolobow, Theodore, 108
Kung, Kuan, 117

Lackey, 122
Lancet, 18
Lappé, Mark, 85
Laughlin, Harry, 90
Lederburg, Joshua, 72
Leeuwenhoek, Antonius van, 99
Leviticus, Book of, 116
Lewis, C. S., 150
Life, beginning of, 14, 17
Lifeboat Ethic, 12
Life or Death, 11
Liley, A. W., 55
Liver extract, 4
Living will, 33
Lowe, Charles, 18
Lower, Richard, 116
Lucy, Cornelius, 106
Lysergic acid diethylamide, 7

Malaria, 47
Maple syrup urine disease, 48
Maryland screening, 78, 82
Massachusetts Supreme Court, 15
McNath, William, 36
Means to preserve life, ordinary and extraordinary, 30, 32, 34; reasonable and unreasonable, 34–35; usual and unusual, 34–35;
Medea, 116
Mendel, Gregor, 41
Mendelian laws of heredity, 90
Merrill, Carl, 59
Metabolic disorders, detection, 55
Mintz, Beatrice, 59
Miscegenation laws, 91
Moniz, Egas, 120
The Moral Decision, 12
Motulsky, A. G., 50
Muller, Herman, 147
Mutagenesis, 70
Mutagens, measuring, 61
Mutations, 61, 65

Nadler, Henry, 55
National Academy of Sciences, 12, 82
National Commission for Protection of Human Subjects of Behavior and Biomedical Research, 121, 126–27, 136, 138
National Foundation—March of Dimes, 45
National Institutes of Health, 5
National Research Act, 125–26
National Resources Defense Council, 72
Natural Death Act, 32
New Jersey Supreme Court, 27
New York screening program, 82
Non-therapeutic research, 127–28
Nucleotide, 61
Nuremberg tribunal, 143

Olds, James, 122
Oral contraceptives, 3, 6–7, 109–11, 154
Organ replacement, 115
Organ transplants, 20, 115, 117
Origin of the Species by Means of Natural Selection, 40
Osler, William, 31
Ovum transplant, 107

Pancoast, W., 99
Patients for experiments, 135
Penicillin, 4–5, 7, 12
People, 14
Samuel Pepys' Diary, 116
Petrucci, Daniele, 104
Pharmaceutical Manufacturer's Association, 138
Phenotype, 97
Phenylalanine, 48, 81
Phenylketone, 48
Phenylketonuria, 48, 50, 57, 75, 79, 82, 133; effect of environment, 49; screen for, 81
Phenylpyruvic acid, 48
Phipps, James, 142
Pincus, Gregory, 109–10
Pius XII, Pope, 30, 34
PKU (see phenylketonuria)
Placebo, 146–47
Planned Parenthood Association, 14, 17
Plato, 40, 53

Polio vaccine, 3, 5–6
Politics, 54
The Population Bomb, 12
Porphyria, 39
Pringsheim, N., 99
Priorities, resources, 156
Prisoners in research, 135–36, 138
Probability of genetic diseases, 43
Progestin, 6
Prokaryote, 59–60, 69
Prometheus, 167
Protein, determined by genes, 41
Psychosurgery, 120–21
Psychotherapy, 119–20
Ptolemy, 40, 53
Pyloric stenosis, 48, 51
Pythagoras, 53

Quality of life, 149
Quinlan, Karen, 27

Raphael, 53
Recombinant DNA, 159, 162–64, 166
Reproductive engineering, 98
Responsibility, of scientists, 154, 160, 164–65; of public, 164
Restriction enzymes, 62
Rh factor, 55, 95
Rhode Island screening, 82
Ribonucleic acid, 41, 58
Right to know, 80, 85
Rights, of fetus, 126; of individuals, 93, 95–96, 149, 151, 156, 158–59; of patients, 32, 111, 159; of scientists, 159; of society, 8, 93, 158
Rindfuss, R.R., 102
Risk vs. benefit in research, 130, 151
RNA (see Ribonucleic acid)
Roberts, Colin, 18
Rock, John, 109–10
Rutter, Wm. J., 63

Salk, Jonas, 5–6
Sanctity of life, 149
Sanger, Margaret, 90
School of Athens, 53
Scientists' accountability, 153; responsibility, 154, 160, 164–65
Searle, G. D. & Co., 110
Seneca, 32
Sex chromatin, 55
Sex determination, 98, 100–102
Sherman, J. K., 99–100
Shettles, Landrum, 104, 107
Siamese twins, 155
Sickle cell disease, 43, 45–46, 49, 57, 74–76, 78, 80
Small pox, 142
Socrates, 40, 53
Somatostatin, 63
Spallanzani, Abbé Lazzaro, 98–99
Species, number of, 40
Sperm separation, 101
Standards, genetic norm, 87
Steptoe, Patrick, 104–105
Sterilization, 90–91, 93
Stilbestrol, 130–133, 144
Sulpha drugs, 3, 7
Superovulators, 103, 111

Tay Sachs' disease, 48, 56, 75, 78
Test tube babies, 102
Thalassemia, 49
Therapeutic research, 127
Thomas, Louis, 72
Thomas, Saint, 11, 126
Tissue culture, 3, 5–6, 59
Tobacco mosaic virus, 58
Tooze, John, 70
Tranquilizers, 3, 5, 7
Transduction, 58
Transferase, 58
Transformation, 58
Transplants, 118
Triage, 12–13
Triethylenemelamine, 61
Trisomy-21, 46
Twins, 18
Tyler, Edward, 99
Tyrosine, 48

U. S. Consumer Products Safety Division, 161
U. S. Court of Custom and Patent Appeals, 73

U. S. Public Health Service, 90
U. S. Supreme Court, 16, 91, 129
U. S. vs Holmes, 11

Vaccines, 7; contraceptive, 101; hepatitus, 141; measles, 142
Valenti, Carlo, 55
Vasectomy, 100
Vatican, 53
Viability of fetus, 16, 19, 129–30
Vicia-faba, 58
Vocational Rehabilitation Act, 95

Walzer, Stanley, 76–77
Watson, J. D., 41
Westminster Community Hospital, 15
Westoff, C. F., 102
William Brown, 11
Willowbrook State School, 141–42

Xeroderma pigmentosum, 58
X-rays, 43
XYY Chromosome, 76–77, 80

Zapol, Warren, 108
Zero growth society, 159
Zeus, 167
Zoroaster, 53